Understanding a Nautical Chart

Understanding a **Nautical Chart**

A Practical Guide to Safe Navigation

Paul Boissier

FERNHURST
BOOKS

Reprinted in 2019 & 2022 by Fernhurst Books Limited

Second edition published in 2018 by Fernhurst Books Limited
The Windmill, Mill Lane, Harbury, Leamington Spa, Warwickshire. CV33 9HP. UK
Tel: +44 (0) 1926 337488 | www.fernhurstbooks.com

First edition published in 2011 by John Wiley & Sons Ltd

This product has been derived in part from material obtained from the UK Hydrographic Office with the permission of the UK Hydrographic Office, Her Majesty's Stationery Office.

© British Crown Copyright, 2018. All rights reserved.

NOTICE: The UK Hydrographic Office (UKHO) and its licensors make no warranties or representations, express or implied, with respect to this product. The UKHO and its licensors have not verified the information within this product or quality assured it.

THIS PRODUCT IS NOT TO BE USED FOR NAVIGATION

Front cover photograph: © Adam Burton, Alamy

A catalogue record for this book is available from the British Library

ISBN 978-1-912177-07-3

Updated by Daniel Stephen
Printed in Czech Republic by Finidr

Contents

C

To Susie

and

to all the wonderful people at the UK Hydrographic Office, who do so much to keep us all safe.

1 What is a Chart?

I have been lost at sea on a number of occasions.

Admittedly, it was before the advent of GPS, and the fault was all mine. I was never very good at astronavigation. But the fact remains that getting lost at sea is quite scary. One bit of the sea is just like another: whichever way you look, it is flat and wet, leading to an inaccessible horizon, with no clear indication of depth and, without a compass or the sun, no sense of direction. When you are lost and out of sight of land, you naturally become a little paranoid, like a traveller in a Siberian snowstorm, wondering whether you will run right into a cliff or a rocky shoal the moment the sun goes down. You can only admire the early explorers, for whom seafaring was a constant battle against the elements, the unknown world and, for all they knew, sirens, sea monsters and mermaids.

But the courage and determination of these early mariners, who opened up the globe and spent years on end surveying its waters and coastline was not wasted. Their work, coupled with advancing technology and developments in the science of hydrography, has made it much easier for the mariner of today. We really should no longer worry about getting lost. But even if you know where you are, your safety still relies heavily on your ability to read a chart and to interpret the data which the chart and its associated publications give you. And it is my purpose in this book to help you stay safe by focusing on how you find and use that information.

So, what then is a chart? It is a remarkably high-quality document, made out of special paper that is designed to hold its shape when wet. It provides, in easily accessible form, a representation of some of the most important data that a mariner needs in that part of the world. In short, it is the special ingredient which turns this (empty sea and an awful lot of sky):

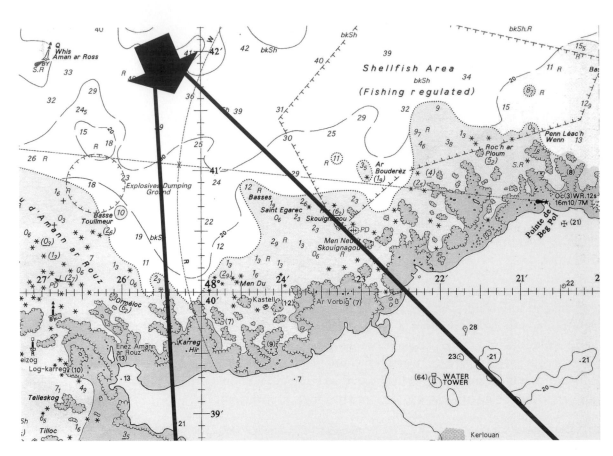

. . . into the chart shown above.

Without a high-quality chart, how else would you know that just two miles to the south-south-east of your position the depth reduces rapidly and becomes dangerously rocky? Or that, if you threaded your way through the rocks and beacons just out of the photograph to the right, you could work your way into an inlet behind the headland? Or, indeed, that just one mile to the south there is an explosives dumping ground where you would be ill-advised to anchor? That's what a chart does for you. It gives you a clear visual representation of your surroundings; it tells you how to get from A to B and, properly used and understood, it helps you to keep out of trouble.

In short, a chart will tell you:

■ where you are

■ what is around you, both under water and on the coast

■ where your destination is in relation to your position

■ and how to get there as safely as possible.

A chart is a very different beast from a land map. A huge amount of effort goes into the recovery of hydrographic information, but the sea is nothing like as well surveyed as the land. And sometimes (we will come to this later in the book), data is deliberately left out when it does not add to the mariner's appreciations of his surroundings in order to improve the clarity of the chart. So you don't have precise little contour lines around every last feature on the seabed. And, of course, on most charts the scale changes from the top of the chart to the bottom – so even reading linear distance needs a little bit of thought.

There are many people who use the seas responsibly – but a fair number too who don't. And right up there on my list of people who could so easily try harder are the ones who set out to sea without an appropriate chart, or with a chart that they haven't bothered to update. Crackers! Our knowledge of the seabed is at best incomplete, and in any case the bottom topography is changing the whole time, through seismic activity, the movement of sediment, human activity and erosion. The least you and I can do is keep up to date with the bits that the hydrographer does know about.

You quite simply cannot stay safe, and look after the safety of people who come to sea with you, unless you have (and are using) a good, accurate and up-to-date chart. And even then there are risks. Good charts may be expensive but they are a lot cheaper than the vessel that you are going to sea in and the expense is far less important than the safety of your crew. So charts really aren't optional. Digital or paper charts, preferably both.

Digital charts are becoming increasingly commonplace in all vessels from a small yacht to a large container ship. I will deal with digital charts in Chapter 17, but in the early part of this book I will talk about paper charts; they are still carried in most boats' and ships' navigation suites, and the great majority of the information – and the way that we interpret that information – is identical.

How to Treat a Paper Chart

When you get your hands on a paper chart, inspect it carefully. It is a well-designed document, the product of years of development, research and painstaking cartography, and it has been printed with immense accuracy. You should always treat it with care.[1] There is an often-told story about a ship one dark night taking a 50-mile detour around a coffee stain on the chart which the watch officer had mistakenly thought to be a reef.

Admiralty charts are designed to take a lot of punishment, but even they have their limits. I can remember keeping a watch on one of the Royal Navy's last open bridge warships in a North Sea storm when the chart became so wet and waterlogged that it was completely impossible to write on it – but even under these conditions it still kept its shape.

What Information Can You Expect a Maritime Chart to Contain?

The extraordinary thing about a chart, when you come to look at it in detail, is just how much diverse information it does contain. Different charts of the same stretch of water, drawn to various scales, will carry subtly different information and I would most strongly recommend that, each time you take out a chart to navigate on, you spend a few minutes studying it as a whole, reading the notes and absorbing the detail. There is a lot to take in.

In general, then, you would expect a chart to contain:

----

[1] See Chapter 5.

1

- a distance scale, latitude and longitude references

- a north reference, both magnetic and true, together with one or more compass roses

- the coastline in serious detail and the hinterland in rather less detail, focusing largely on the features most likely to be of interest to the navigator

- depth information with relevant contours and intelligently selected soundings

- anchorages

- underwater dangers, including rocks, wrecks, overfalls and obstructions

- outline tidal height information

- lights and navigation marks, sound signals, buoys, transits and shipping lanes

- fishing areas, energy platforms, separation lanes, international boundaries

- survey information showing the date and the thoroughness of surveys of each section of the chart

- a list of applied corrections and their dates

- and an awful lot more that we will come to later in the book.

Colour Convention of Charts

The colour convention of charts is important as well. From 1800 until 1968, all UK Hydrographic Office (UKHO) charts were published in fathoms and feet, but since 1968 there has been a gradual conversion of UK charts to metric data. The great majority of charts published by the UKHO are now in metric format, that is to say that depths and heights are shown in metres. These charts are printed in the colours that you will be most familiar with: the land shaded a kind of buff yellow and the sea and shoreline variously green, blue and white according to depth. There remain a number of charts, however, where depth and height are measured in imperial units, specifically charts published by the US Hydrographer and republished in the United Kingdom, and a number of less-used charts, generally of waters remote to the United Kingdom, which have yet to be converted to metric scales. These charts are printed in black and white with a number of coloured additions to identify specific features, like magenta flashes to highlight navigation buoys. More recent charts also use blue shading to indicate shoal water.

Digital Charts

I will be going into digital charts in more depth in Chapter 17, but it is important to recognise that they are pretty much the same as paper charts, drawn from the same data and using the same conventions. Digital charts are, however, produced in two separate and distinct formats, which it is important to both understand and recognise:

- **Raster Charts:** Published by the UKHO under the title of Admiralty Raster Chart Service, or ARCS for short. These are straightforward electronic copies of the equivalent paper chart, directly scanned onto a CD. There is nothing added or taken away. As such, they are less versatile than the multi-layered Vector chart, but safer too because you cannot 'lose' information by having

1

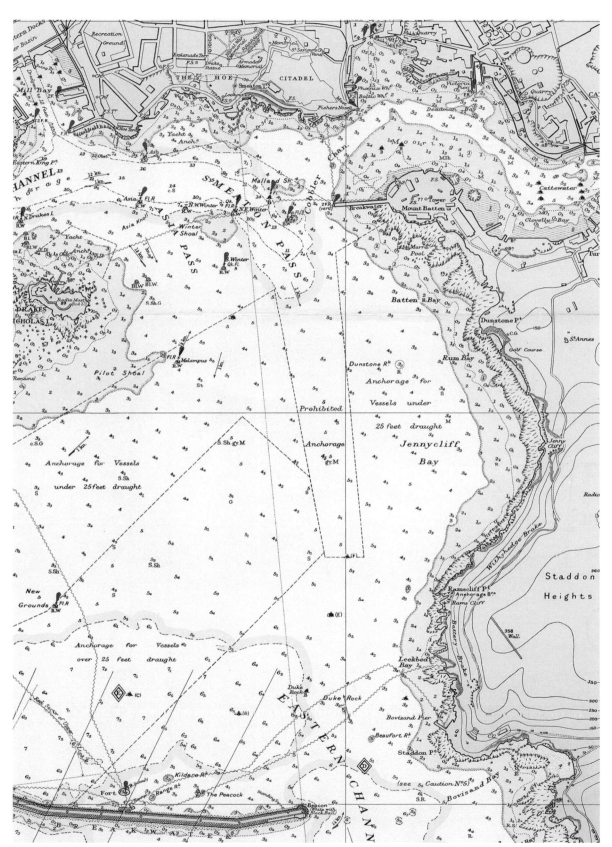

Imperial Chart of Plymouth Sound Published in the 1960s

1

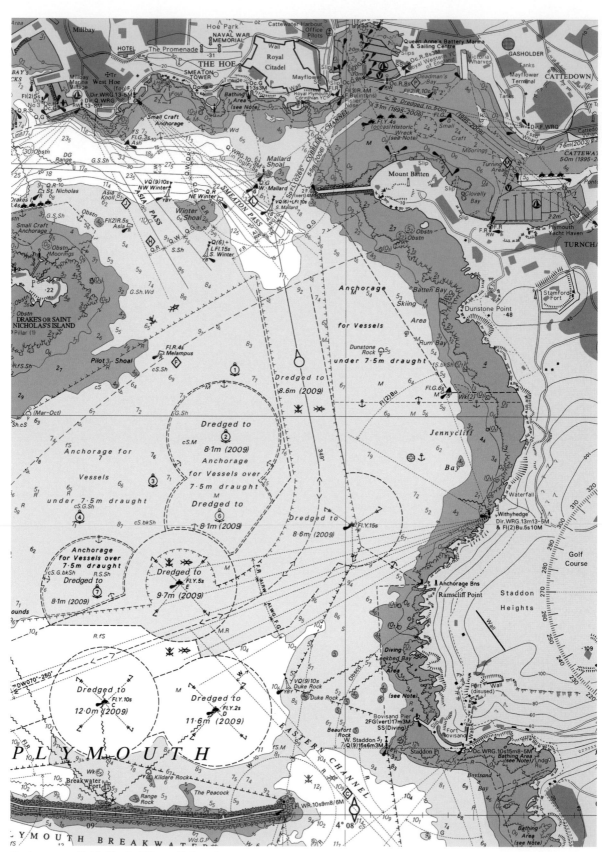

Contemporary Metric Chart of Plymouth Sound

a layer of information inadvertently switched off. The Raster chart is quite simply the digital equivalent of a paper chart.

■ **Vector Charts:** These, too, are published by the UKHO as Electronic Navigational Charts, or ENCs, with an expanding worldwide coverage. They are also produced by a number of other charting authorities, to varying specifications. A vector chart is essentially a blank screen onto which a number of 'layers' of information are added: the coastline, land features, soundings, titles, currents, tides and tidal heights, etc. In fact, there is no limit to the number of layers that you can add, providing the mariner with AIS (automatic identification system), radar, harbour entry information, aerial views and much more as he or she requires. But the danger of this is that it all adds 'clutter' to the screen so, at any one time, a number of the layers need to be switched off in order to make the chart usable. So you never quite know what you are missing. Many mariners like to use a paper chart (or its raster equivalent) alongside a vector product so that they can see the general picture in a familiar format on the chart, with all the information in its allotted space, before looking more closely at the details on the vector display.

1

2 A Chart is Never 100% Accurate

You should treat any chart with a degree of healthy suspicion.

I don't know about you, but my first instinct when someone tells me something more or less credible with sufficient conviction is to believe them, particularly if that person is someone I trust. And it's just the same when you see something on a chart, especially an Admiralty chart, which is after all a sort of archbishop among charts in terms of its pedigree and credibility. We all naturally tend to believe what we see on a chart without questioning its reliability . . . and that's a dangerous thing to do.

Don't get me wrong: in my opinion Admiralty charts are still the best, most accurate and most reliable charts available, but even the 'archbishop' is, by his own admission[1], less than 100% accurate – and as a mariner you need to understand how and when to make allowance for these inaccuracies.

There are a lot of reasons for this: many charts contain areas of old surveys, sometimes even centuries old, because more modern data is just not available. There are two problems with old survey data: firstly, navigation techniques were not as good in the past as they are today, so the positioning of features on a chart may be suspect; secondly, the technology of surveying was less advanced, and old surveys inevitably left gaps in the survey coverage in which even quite sizeable dangers could be lurking. New and more thorough surveys regularly reveal shallow patches, previously unknown, which reach up to within a few metres of the surface.

But there are other reasons to question the accuracy of information that you are getting from your chart and its associated publications: the seabed may have shifted since the most recent survey; the chart may not be up to date; and you may from time to time experience tidal surges caused by strong winds and atmospheric pressure variations, which have the power to change the depth of water dramatically. There are also places where the seabed is just too mobile – with strong tides and a soft seabed – and the hydrographer simply does not bother charting the area at all. There are some examples of this very close to home. Take a look at Chart 1346 (opposite) of the Solway Firth, close to Carlisle on the western end of the border between Scotland and England. There is a section, about 10 x 15 nautical miles, where the seabed is so unstable that the hydrographer does not bother to chart it at all – and that area of the chart is left white.

Of course, it's not as bad as it sounds. Many old surveys are surprisingly accurate, and quite good enough for the shipping that passes that way. Matthew Flinders' survey of South Australia conducted in 1801–1802, for instance, was so accurate that it is still the basis of Admiralty charts

---

[1] *The Mariner's Handbook*: 'While every effort is made to ensure the accuracy of the information on charts, information may not always be complete, up to date, or positioned to modern surveying standards. No chart is infallible. Every chart is liable to be incomplete, either through the imperfections in the survey on which it is based, or through subsequent alterations to the topography or sea floor. However, in the vicinity of recognised shipping lanes, charts may be used with confidence for normal navigational needs. The mariner must be the final judge of the reliance he can place on the information given, bearing in mind the particular circumstances, safe and prudent navigation, local pilotage guidance, reference to source information, and the judicious use of available aids to navigation.

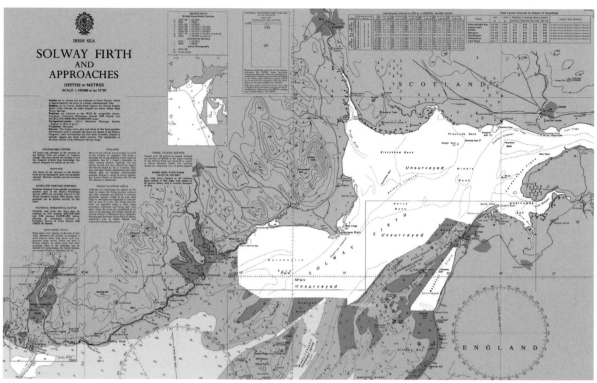

Chart 1346 – Solway Firth

of the area. In any case, the chart compiler generally tends to err on the side of caution to give you and me that extra margin of safety to keep us from hitting the bottom. But even so, it pays to treat the information very carefully.

On 7 August 1992, the *QE2* grounded on a shoal off the coast of Martha's Vineyard on the eastern seaboard of the United States. Her draught at the time was 32 ft 4 in, the height of tide was about 1 ft 6 in, and the charted depth of water at the point where she grounded was 39 ft. By straightforward calculations she would have had about 8 ft of water under her keel. At the time she grounded, she was travelling at a speed of 25 knots. It transpired that the most recent survey of the area had been conducted in 1939 using an echo sounder. A post-event side scan survey, however, showed that the shallowest depth of water in the area was 31 ft and that the depth of water at the precise point where she grounded was 33 ft below Chart Datum, 6 ft less than the charted depth. It is likely that this patch of shallow water had quite simply fallen into one of the gaps left by the earlier echo sounder survey. A ship moving fast in any shallow water is certain to squat[2] a foot or so . . . and suddenly, in the blink of an eye, you have reduced a confidently predicted clearance of 8 ft under the keel into an agonising incident with an awful lot of paperwork to fill in.

As a mariner, you need to know how to look at a chart in order to assess the risks and inaccuracies that it contains, so that you use it intelligently, recognising its limitations and identifying where you have to navigate with more caution. The information is often right there on the chart, paper

---

[2]'Squat' is the technical term for an increase in a ship's draught caused by the flow of water between the ship's bottom and the seabed. It increases significantly at high speed and in shallow water.

2

or digital. It's just a matter of knowing where to look, how to interpret the information and how much of a margin of error to apply.

Survey Data

Within the United Kingdom, serious offshore surveying started in about the middle of the eighteenth century with the voyages of Captain Cook, and has continued ever since, progressively improving the quality, coverage and accuracy of charts. The majority of surveys undertaken before 1935 were conducted with visual fixing and a leadline. From the mid-1930s, a series of electronic navigation aids (LORAN, Decca, satnav, etc.) made offshore position-keeping increasingly accurate, leading to the advent of GPS and Differential GPS (DGPS) in about 1980. The creation of portable computers followed by an exponential growth in computer power and the development of ever more advanced survey techniques have all served to massively improve the quality and reliability of our charts. This means that surveys which have taken place over the last quarter of a century are a quantum leap ahead of anything that has gone before, with greater reliability and fewer gaps in the coverage. Even so, it is a painstakingly slow process to get an accurate representation of the seafloor, and the great majority of the oceans remains thinly surveyed.

According to Dr Jon Copley of Southampton University:

> The entire ocean floor has now been mapped at up to ~5 km resolution, which means we can see most features larger than ~5 km across in those maps..... Multi-beam sonar systems aboard ships can map the ocean floor at ~100 m resolution, but only in a track below the ship. Those more detailed maps now cover about 10 to 15 percent of the oceans, which is an area roughly equivalent to Africa in size.

This means that, along with older, single-beam echosounder data from ships crossing the ocean we probably have 'soundings' for a total of something like 20% of the ocean floor. These percentages don't change much from year to year. Around the coast of the United Kingdom, it is surprising how much of the seabed has not been thoroughly surveyed – even now.

Take Lundy Island in the Bristol Channel, for instance. Lying only about 15 miles offshore, close to the English and Welsh coasts, this is not a particularly remote, inaccessible or difficult area to survey. But have a look at these three charts:

The first is a simple scan from the largest-scale Admiralty chart of the north-west corner of Lundy Island, showing an area of uneven bottom and overfalls just off the coast. This data was plotted from a

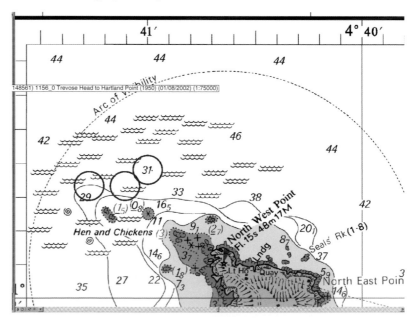

leadline survey made in 1879, and this is the most accurate chart of the area that you could have bought up to about the summer of 2009.

During the early part of 2009, however, the UKHO commissioned a new survey of the area, using multi-beam sonar, which revealed a number of areas of shallow water that had never previously been identified. A detail from the new survey is shown in the second diagram. You will also see how very much more data is gathered by a modern survey; even this is only a small percentage of the topographical data that was recovered, but it gives the hydrographer a much more accurate representation of the shape of the seabed from which to draw the chart, and much more confidence that he has picked up all of the potential dangers.

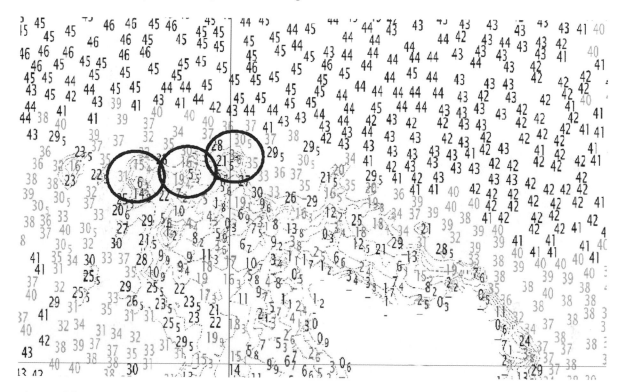

The problem with using old data is best illustrated by superimposing the new survey onto the old chart, and this is shown, in a magnified format, in the third diagram. You will see how, in the most north-easterly circle, the original survey shows a depth of 31 metres, which is accurate enough with regard to the general depth of water in the area. However, the original leadline survey completely missed a pinnacle of 3.6 metres situated almost directly below this sounding point. The other two points that I have circled show depths of 5.5 metres and 6.2 metres respectively in an area that, according to the chart, you would expect something between 20 and 30 metres. If you look closely, you will see a number of other inconsistencies.[3]

I will probably mention it a dozen times over the course of this book, but this is a graphic illustration of why it is absolutely vital: **you must keep your charts up to date**. You could have decided to take your ship for a close pass of Lundy Island in the summer of 2010 using uncorrected charts, blissfully

---

[3] I am indebted to Mr Roger Cavill, Head of Regional Team 1A at the UKHO for these images.

unaware of the new data which was there in the *Notices to Mariners*, sitting in your chart room drawer just waiting to be drawn in . . .

And just to show you what it really looks like, below is an example of the way in which this massive richness of multi-beam survey data can be displayed, showing the seabed in extraordinary clarity and accuracy. This is exactly the same stretch of water, to the north of Lundy Island, seen from the north-west in a three-dimensional view. The tip of the island is shown in the top left-hand corner of the picture, with the seabed in the foreground, all displayed in glorious Technicolor. The shallowest patches are coloured red and the deepest are shown in blue, and you will see just how many pinnacles there actually are in this stretch of water.

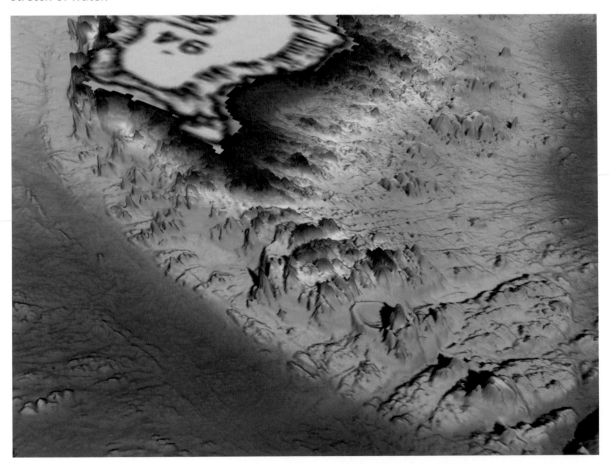

It is really no surprise that the early surveyors, equipped only with a leadline, failed to pick up anything but the most superficial detail from this complex bit of the seabed.

There are still a number of parts of the British coast, let alone other parts of the world, that remain poorly surveyed. This next chart – Q6090, dated March 2017 – shows just how much of the United Kingdom's waters have yet to be thoroughly surveyed.

Only the dark green patches have been surveyed to the latest standards of accuracy by swathe bathymetry, which provides the most detailed picture of the sea floor. The yellow patches have been surveyed with a single-beam echo sounder, which only identifies the depth (and obstructions) along the ship's track. The red areas, of which there are many close inshore, have only been surveyed with lead line, or may even be unsurveyed. Yachtsmen in particular may like to check out the survey dates on their charts of the Channel Islands: much of the area was surveyed well over 100 years ago.

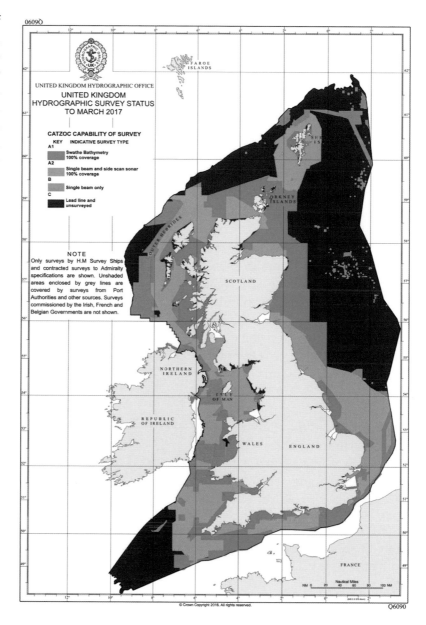

There are undoubtedly more surprises out there around the coast that are just waiting to be discovered and, if you venture into waters that might be less well used, or poorly surveyed, you must be aware that your chart might not be as reliable as you would like to believe.

It's easy to overemphasise the dangers of chart inaccuracy. Of course, the navigator needs to be constantly aware of the limitations of the chart in use, but the danger is likely to be very much less when operating in areas of frequent and regular maritime activity, for the simple reason that other people will have trodden your path before you. When you venture off-piste, however, you need to be more careful.

Chart Coverage and Consistency

Before you use a chart, you should check the accuracy of the source data used by the chart compiler when drawing up the chart. And the place to look is on the chart itself in the margins, where you will nearly always find a 'SOURCE DATA' diagram and table.[4] The diagram takes the form of a scaled replica of the chart divided into discrete areas, each of which represents an individual packet of survey data.

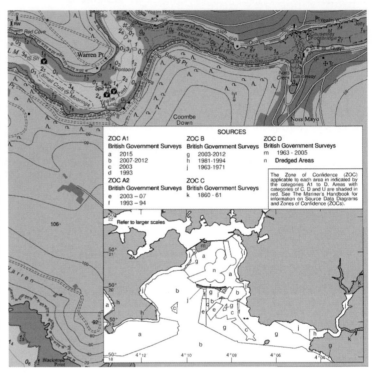

Since 2017, the UKHO has described the accuracy of a chart survey by Zones of Confidence, or 'ZOC', which describes the accuracy and coverage of the sea floor information that is shown on the chart. This is the system the Hydrographic Office uses on its digital charts, so it makes sense to use the same system their paper charts. I will discuss ZOCs further in Chapter 17, but you can see that here, on Chart 30 – Plymouth Sound and Approaches – some of the survey data around Drakes Island at the top of the Sound, and on the upper reaches of the River Yealm, is rather less than reliable. Most of the Sound, however, is regularly used by a variety of shipping, and it has been surveyed to a very high standard.

It is important to realise that the 'ZOC' codes refer solely to the accuracy of the survey; in areas where the bottom is unstable, the survey may be accurate, but it is possible that some changes in depth may have occurred since the survey was conducted – so you still need to treat the information with caution.

Zone of Confidence code	Position accuracy of bottom features	Depth accuracy	Sea floor coverage
A1	5m	0.5m +1% depth	Full area search
		0.8m accuracy at 30m depth	Significant sea floor features detected and depths measured
A2	20m	1.0m + 2% depth	Full area search
		1.6m accuracy at 30m depth	Significant sea floor features detected and depths measured

---

[4]This applies to Admiralty charts and the charts produced by many other national authorities. Some commercial charts, particularly those designed for leisure users, do not print a source data diagram.

B	50m	1.0m + 2% depth 1.6m accuracy at 30m depth	Systematic survey Some uncharted hazardous features may exist but are not expected
C	500m	2.0m + 5% depth 3.5m accuracy at 30m depth	Depth anomalies may be expected
D	Worse than ZOC C	Worse than ZOC C	
U	Unassessed. The quality of the data has yet to be assessed; thus other means are required to ascertain survey quality		

Charts published before the summer of 2017 have the older-style source data diagrams, like the illustration below:

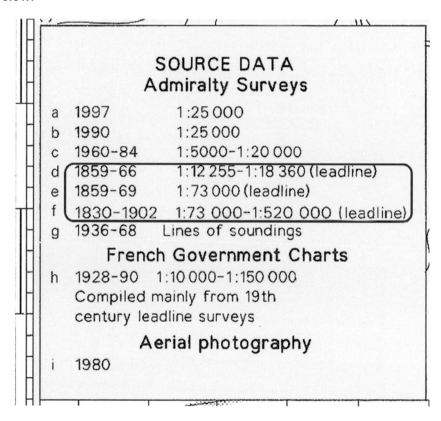

This example of the older style source data table is taken from a pre 2017 chart of the Channel Islands. The first thing that will strike you is that much of the data originates from surveys that took place in the mid-nineteenth century, and that only a relatively small area has been surveyed by anything other than leadline. This does not necessarily make the data less accurate; much of the early surveying work is surprisingly accurate, but there are almost certainly gaps in the coverage and, as you navigate around this area, you may wish to ask yourself whether, in the interests of responsible seamanship, you may want to increase your margins of error a fraction . . . just in case.

If you look at the source data diagram, you will see a 'scale' column, containing entries like '1:25 000'. Not a lot of people really understand what this means. This column allows you to get a feel for the survey line spacing. The assumption is that a surveyor will draw his survey lines on a working chart with a spacing of 5 mm. So, a 1:20 000 survey will be derived from a survey where the lines are 20 000 x 5 mm (or 100 metres) apart. This is the separation between adjacent tracks of the survey vessel which, in the days before sidescan sonar, was the maximum size of an uncharted obstacle that the surveyor could have missed. The area marked **f** on this chart has a scale of 1:520 000, which indicates that, in the deeper water to the west of the Casquets, some of the leadline survey runs were separated by 2 600 metres – substantially more than a mile. You will note, moreover, that this particular survey was conducted in the latter half of the nineteenth century.

The most recent surveys, carried out with multi-beam (or swathe) survey systems quite simply do not leave gaps and the term 'full seafloor coverage' is used instead of a scale.

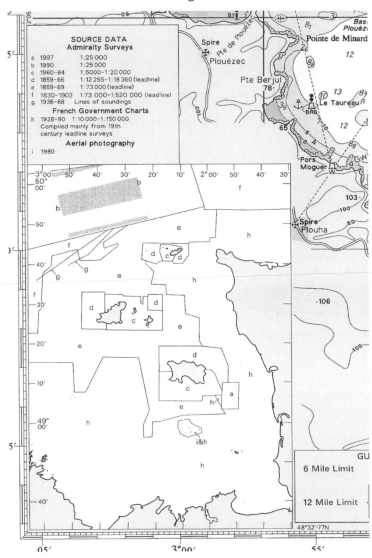

Source data is also displayed on Admiralty digital charts – as a serious navigator, you would want to ask some pretty tough questions of any chart that does not display this information. Raster digital charts display Source Data in exactly the same way as a paper chart, but on a vector digital chart there is a specific layer called 'CATZOC' or 'Category of Zone of Confidence', which will give you a 0- to 6-star rating for the quality of information at any point on the chart. I will look more closely at this in Chapter 17.

This useful table is extracted from an earlier edition of *The Mariner's Handbook* and gives some indication of the increasing reliability of charts over the last 150 years.

These advances in cartography are graphically illustrated by the following three charts of the Isle of Wight, which were drawn in 1685, 1852 and 2006, respectively. These are followed by a fourth chart, which uses the 2006 data to depict the bottom contours in great detail.

Date	Sounding Method	Fixing Method	Remarks
Pre-1865	Leadline	Angles to local landmarks	Surveys were mainly concerned with recording previously undiscovered lands. More attention was given to fixing the coast than to providing sounding. Soundings, where present at all, tend to be sparse with irregular gaps between them. The quoted scale is largely irrelevant when used to judge likely sounding density.
1865	Leadline	Angles to local landmarks	Steam replaced sail in British survey ships and regular lines of soundings begin to appear. The scale of survey will give an indication of the expected density of soundings. Inshore, where boats were used instead of ships, oars remained the motive power and sounding lines continued to be irregular.
1905	Leadline	Angles to local landmarks	Steam replaced oars as the power for survey boats, allowing regular lines to be extended to all areas and water depths of the survey. The scale of survey gives an indication of the expected density of soundings.
1935	Single-beam echo sounder	Angles to local landmarks	Greater ease of collecting soundings allowed far denser surveys to be gathered. The scale of survey gives an indication of the expected density of soundings.
1950	Single-beam echo sounder	Electronic position fixing	Greater accuracy and consistency of position fixing extending further offshore than was possible with angles to shore marks.

1973	Single-beam echo sounder and sidescan sonar	Electronic position fixing	Sidescan sonar allows surveyors to locate hazards that exist between survey lines. For the first time the surveyor will have covered the entire seafloor.
1985	Single-beam echo sounder and sidescan sonar	Satellite position fixing	Introduction of satellite positioning allows the surveyor to accurately position his ship anywhere in the world to a common datum.
2000	Swathe echo sounder	Satellite position fixing	Swathe systems replace single-beam echo sounder and sidescan sonar. Multi-beam not only allows the surveyor to detect obstructions between survey lines but also allows depths to be gathered over them.

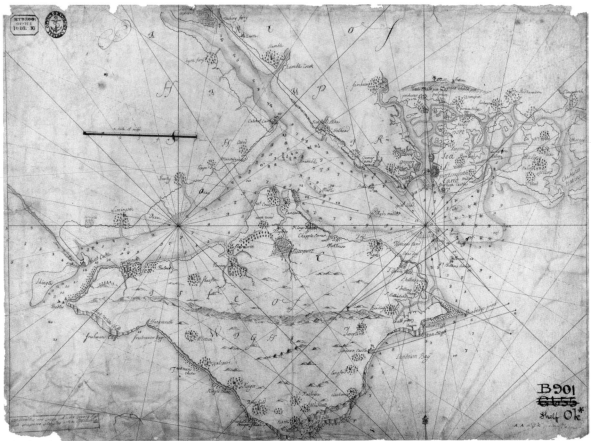

Isle of Wight 1685

2

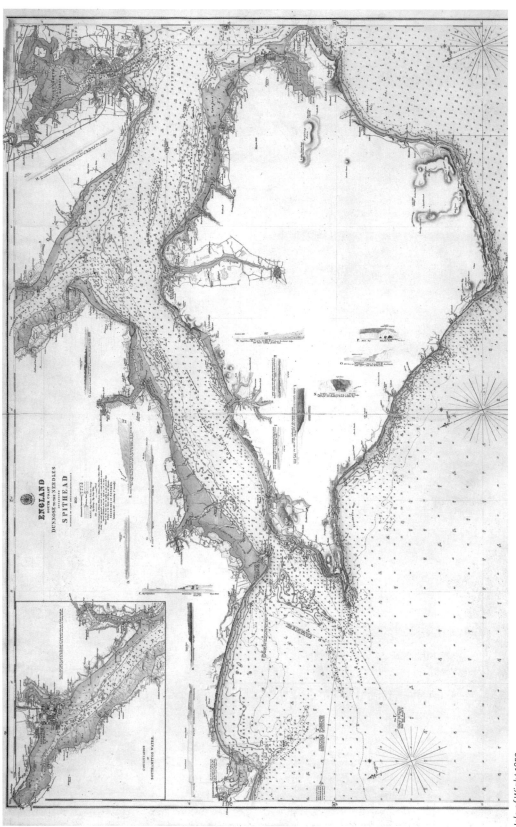

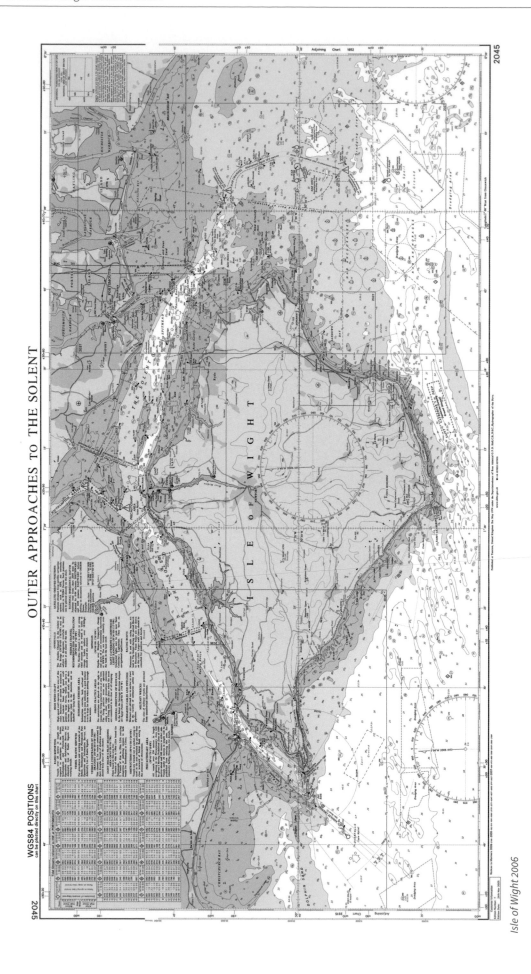

OUTER APPROACHES TO THE SOLENT

Isle of Wight 2006

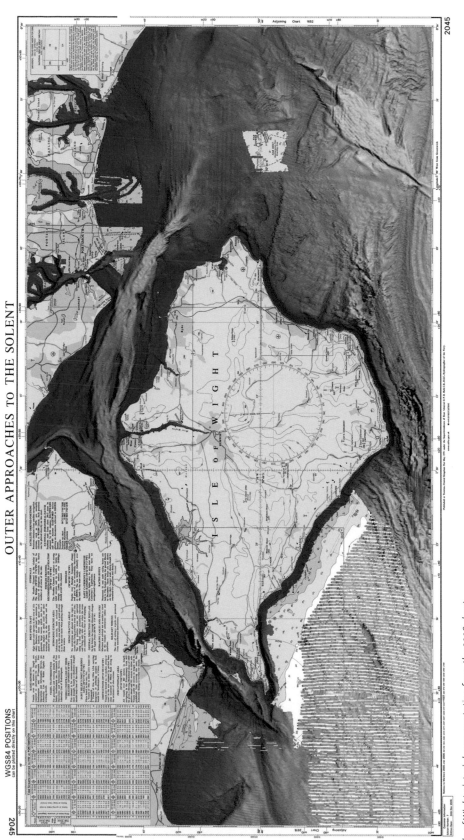

Isle of Wight pictorial representation from the 2006 chart

2

Horizontal Datums

There was a time when some people were absolutely convinced that the world was flat. At school in the 1960s, I was taught that the world was a sphere – a great blue-and-green beach ball in orbit round the sun. With the advent of GPS and modern surveying techniques we have discovered that the world is not in fact a sphere but an 'oblate spheroid' (a handy term for pub quizzes) or, to you and me, a sphere which has been flattened at the poles and expanded at the equator by the centrifugal force of the earth's rotation.[5] This is clearly a more complicated shape to plot on a chart than a perfect sphere or a flat earth, and there have been a number of attempts to create virtual models of the shape, principally to allow GPS satellites, in orbit around the earth, to provide us with an accurate position on the earth's surface.

The most commonly used model is the 'World Geodetic Survey 1984' abbreviated to 'WGS84', and critically it is this model that is used by the US Global Positioning System (GPS). 77% of Admiralty charts are now referred to the WGS84 datum, but many are still plotted to a variety of older datums, which may not be so accurate, or so GPS-friendly. Accordingly, whenever you are using GPS, you should always check the horizontal datum of your chart (in the main Title Block) to see if there are any corrections that you need to apply.

If you want to plot a GPS position to any non-WGS84 chart, you will have to apply a correction, and any chart that is not drawn to the WGS84 datum will carry a conversion table and a worked example (see the illustration) that allows you to adjust from one datum to another.

Chart Datum; all other heights are above Mean High Water Springs.
Positions are referred to European Datum(1950) (see SATELLITE DERIVED-POSITIONS note).
Navigational marks: IALA Maritime Buoyage System—Region A (Red to port).
Projection: Mercator

SATELLITE–DERIVED POSITIONS

Positions obtained from satellite navigation systems, such as the Global Positioning System (GPS), are normally referred to the World Geodetic System 1984 Datum. Such positions must be adjusted by 0·06 minutes NORTHWARD and 0·08 minutes EASTWARD before plotting on this chart
Example:
Satellite–Derived Position
(WGS 84 Datum)　49°05′·50N, 002°15′·50W
lat/long adjustments　0′·06N　　　0′·08E
Adjusted Position 49°05′·56N, 002°15′42W
(compatible with chart datum)

These cross-datum errors can be considerable: the largest-known discrepancy between the charted position and the satellite-derived position is a massive 7 miles in the middle of the Pacific Ocean. Which means that, without applying these corrections, you could miss that delicious coral atoll altogether!

Most digital charts have been either drawn up to WGS84 or adjusted to remove the errors.

---

[5]Another good fact for quiz night: the distance between poles is about 42.7 kilometres less than the equatorial diameter. Who can say that navigation isn't fascinating?

2

Soundings and Vertical Datums

The other datum used in chart-making is the 'vertical datum', from which all depths of water are measured. In the United Kingdom, the charts are referenced against a 'chart datum', which is fixed at a level below the range of normal tides, close to the level of the 'Lowest Astronomical Tide' (LAT)[6] which means that there will almost always be more water than indicated on the chart. But this is not always the case. Other countries use different conventions on their charts.[7] Once a Chart Datum has been established in an area, the hydrographer will use it as his principal vertical datum against which both the height of tide and the sounding data are measured.

Total depth of water:

Soundings are the distance of the seabed or a bottom feature **below chart datum**, and **the height of tide** is the height of the water surface **above chart datum**

So:

Total depth of water = sounding + height of tide

This makes it quite simple to work out how much water you will have at any one time. The TOTAL **DEPTH OF WATER** EQUALS THE **SOUNDING** (distance from Chart Datum to the seabed) PLUS **THE HEIGHT OF THE TIDE** (predicted distance from Chart Datum to the water surface).

But life is never that simple. The soundings on a chart may not always be accurate, as we have already established, particularly if you are using a chart with old survey data. Also, where the seabed is relatively unstable, depths may change rapidly – sometimes by as much as a couple of metres in a matter of months and with this pace of change there is no way that, even with a recently corrected chart, you should trust it to be accurate.

The tidal predictions are highly reliable in normal conditions, but heavy weather can occasionally have a significant effect. The tide tables have been calculated for average barometric pressure, but a sustained difference of just 34 millibars in pressure can cause a difference in tidal height of 0.3 metres (about one foot). The lower the pressure, the higher the sea level.

In addition, the tide is often affected by strong winds; the surface water is blown downwind until it meets an obstruction, where it holds up. A strong northerly wind in the North Sea, for instance, can cause some spectacular (and very dangerous) surges. A particularly severe storm of 31 January 1953 raised sea levels on the east coast of England and the Netherlands by 2.5–3 metres. Negative surges, where the depth of water is less than predicted, can also be caused by weather conditions. On one occasion, in 1982, the depth of water reported in the Thames Estuary was 2.3 metres below the prediction. There is a page in the *Annual Summary of Notices to Mariners* (see Chapter 4) that

---

[6]Lowest Astronomical Tide is the lowest predictable level to which the tide will fall. Other factors, principally meteorological, can exceptionally cause it to fall below this level.

[7]For instance, US charts are referenced to mean lower low water (MLLW), which is higher than LAT. MLLW takes the lower of the two daily low water levels, records them and averages this height over a period of 19 years.

gives important information about the radio warnings that are available to mariners in the North Sea, the Straits of Dover and the Thames Estuary in advance of tidal surges.

So, although the hydrographer does his best, and his predictions are spot on about 99% of the time, his predictions cannot be 100% accurate because there are just too many things going on at sea. Keep an eye on the conditions, pay attention to your echo sounder and try to correlate the observed soundings with your calculations: when the actual soundings and your predictions don't match up, either you have got your calculations wrong, or else your information is inaccurate. Either way, you may be standing into danger.

Is the Chart Up to Date?

There is almost nothing in the world more endlessly painstaking than correcting a folio of charts.

It is detailed, fiddly and you know full well that you will never go near half the corrections that you make. But this work really has to be done regularly, and an uncorrected chat is one that, sooner or later, will let you down. I will talk about how to correct charts in Chapter 5, but before you use a chart, you should always check its current change state, which is recorded on the bottom left-hand corner. You should check the change state against the most recent *Notices to Mariners*. Helpfully, the Hydrographer now prints a QR Code in this corner, which takes you directly to a page on the UKHO website where you can see the most recent change state, from where you can click across to see the relevant notices. You may wish to scan the QR code in this illustration to see what it looks like.

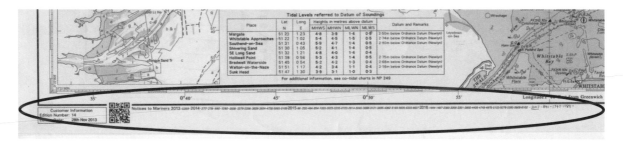

Pre-Flight Checks for your Chart

Take a bit of time over your chart when you first dig it out. It is actually a thing of great functional beauty, and your safety depends on how you use it – so on both scores it merits a few moments of your time. Here are a few things that I would check before I even pick up a pencil:

1. Is it the right scale for the job? Pilotage, coastal and passage charts will often be drawn to different scales and in general you need the largest scale chart to do the job without having to endlessly change from one to another. (Remember the old mnemonic: 'Large scale, small area. Small scale, large area.')

2. Is the chart up to date? Check on the bottom margin to find out when the most recent correction was applied: does that sound reasonable, or should you check either the UKHO website or the *Cumulative List of Notices to Mariners*, both of which contain a comprehensive list of all corrections relevant to a particular chart?

3. Look at the source data table. How reliable is the information you are getting? Also, check for consistency of soundings and unbroken contour lines. Irregular soundings or broken contours are a sign of poor survey coverage.

4. Read the warnings in the margin of the chart: nothing is printed there unless it is important to the navigator.

5. What horizontal datum is being used? Do you need to convert the GPS positions into another datum to plot them on the chart?

6. Read the *Sailing Directions* or local pilot. This is designed to supplement the visual information that you get from the chart. It is only if you do this – a pleasant task, even if a little time-consuming – that you will be able to properly assess the risks of navigation.

3 What the Chart-Maker Does for Us

The chart-maker's job is to look after you and, as far as he can, to stop you from hitting the bottom. The sight of submarine K4 high and dry in Morecambe Bay, taken in about 1913, may not have caused the cartographers of the day to resign en masse, but it would undoubtedly have given them reason to dig out the chart in question and quietly examine the published material. Just to be sure.

When you walk around the UK Hydrographic Office (UKHO) in Taunton, you realise that every day they are making critical decisions about what to include in a chart and what to leave out. This is a highly skilled job. You come away with the unmistakeable feeling that the men and women who draw our charts really do care about getting it right, and that they have a deep-rooted sense of responsibility for our safety out on the water.

Over the years, they have quietly developed a number of very sensible features which they build into charts so that the information that you get always errs on the side of safety and clarity – unnecessary detail is quite simply removed – and the result is a very functional and intelligently designed sheet of paper. This is a list of some of the fail-safe conventions that are built into your chart:

■ The chart-maker recognises that there will be times on the bridge of a ship, or the cockpit of a yacht, when you are under pressure, so the chart is kept as uncluttered as possible – to give you only the information that you actually require at a glance. Small-scale and passage-planning charts contain enough detail to keep you safe on an offshore passage, but not a lot more. Coastal charts show the lights, buoys and navigation features that you need for safe coastal navigation, and it is only when you come down to the large-scale

3

(pilotage) charts that you find all the safety-critical features: significant depths, dangers and aids to navigation. Even then, considerable efforts are made to keep the chart as easy to read as possible.

■ Only a very few of the available soundings, selected from a cast of thousands, actually make it onto the chart. This is a complex and skilled task – making sure that the mariner sees the soundings that he needs for safe navigation, while removing all the others in the interests of clarity. The cartographer's aim is to give you a sense of the bottom topography, with representative and minimum soundings, but without making you have to fight for the information.

■ Depths are always referred to 'Chart Datum' which, in British charts, is approximately the level of the lowest astronomical tide (LAT) in that area. As its name suggests, LAT is the lowest predictable level that the tide will reach under normal circumstances. Accordingly, there will nearly always be water above Chart Datum (check it out in the Tide Tables), so the actual depth of water is unlikely to be less than shown on the chart. So, if you happen to be an intensely idle mariner who can't be bothered to work out the height of tide from the Tide Tables, you will nearly always be safer than the depths on the chart indicate.[1]

■ By the same token, the clearance of overhead obstructions (bridges, power cables, etc.) is now generally referenced to the level of the highest astronomical tide (HAT) to give you a greater air gap than shown on the chart. You do, however, need to check the chart's Title Block to make sure that HAT is being used on that specific chart because, until recently, overhead obstructions were referenced against the slightly lower datum of mean high water springs (MHWS), and older charts may still be calibrated in that way (see the diagram on page 71).

■ A chart will only give you details of shore features that, in the mind of the cartographer, might be useful to a mariner: much of the detail that is not visible just won't be included. So towns are shown in outline, not detail. Hills are marked in broadly representative contour, although the peaks, if they can be used for visual fixing, are accurately pinpointed. The depiction of roads is often rudimentary. Any landmark that can be used for navigation, however, is shown in clear and accurate detail.

■ Metric charts are colour-coded to make them easy to read in a hurry. Dry land (i.e. land that is above the level of MHWS) is depicted in yellow. Areas of the seabed that periodically dry out with the rise and fall of tide (i.e. those which lie between the levels of MHWS and Chart Datum) are shown in green, and shallow water is shown in one or more shades of blue, depending on the scale of the chart and the nature of the seabed. Water deeper than 10 or 20 metres, again depending on the chart's scale, is shown in white. Imperial charts are also usually colour-coded: grey for the land, with equivalent shades of blue for shallow water. These colours are immensely important: they provide an immediate visual signal for mariners to help them recognise when they are standing into danger.

■ Latitude and longitude scales vary from chart to chart with the simple aim of making the scale easier to read for the mariner. Adapting the scale to the chart certainly does help for

---

[1]Don't bank on it though! The height of tide *can* fall below Chart Datum, so in shallow water I would ALWAYS recommend calculating the height of tide, and working out the depth of water properly. Also, charts drawn from foreign surveys, even if sold as Admiralty charts, will always use the datum against which the chart was drawn, which may be higher than LAT.

3

much of the time, but it can sometimes be a trap for the unwary when you are changing from one chart to another and the latitude and longitude scales are graduated differently on the new chart: more than once, this has caused me to plot my position incorrectly. As a result, I always check and double-check a position when changing charts. And I take *real care* if I am crossing the Greenwich meridian or the equator, when the scale reverses: it is incredibly easy to misplot a position.

■ Wrecks are generally marked, like rocks, with the sounding of their highest point. Helpfully, though, the hydrographer also decides whether an underwater obstruction could be a danger to surface navigation and, if it could be considered dangerous, he places a ring of dots (a 'necklace') round the chart symbol to help it stand out from the rest. The criterion for a **dangerous wreck** is linked to the draught of an average merchant vessel, but annoyingly

Depth at which a wreck is no longer considered 'dangerous'	
Date	*Depth criterion*
Before 1960	14.6 metres
1960–1963	18.3 metres
1963–1968	20.1 metres
1968 onwards	28.0 metres

the draught of the average merchant ship has increased substantially over recent years, so the criterion of a wreck that may be 'dangerous to surface navigation' has changed in parallel (see the table). The sensible navigator will certainly use the 'necklace' as a guide, but always check the charted sounding of a wreck that lies close to his track.

■ Rocks have no such depth convention for determining whether they are dangerous to navigation – they are marked as dangerous at the hydrographer's discretion. More detail on rocks and wrecks can be found in Chapter 10.

■ Finally, when a navigational warning is received in the UKHO, the cartographer responsible always assesses the urgency of that information from the perspective of the mariner and processes it accordingly – so that you are only warned by radio or NAVTEX of the most pressing or dangerous problems. The remainder will be quietly compiled and sent to you with the *Weekly Notices to Mariners*, and the least pressing will be held back until the next edition of the chart is published.

The point is that everything is done to reduce the level of risk associated with the use of the chart, and whenever the hydrographer considers that a residual risk does exist, he is duty-bound to tell you either in the margins of the chart itself or, more likely, in the *Sailing Directions*.

In the chart of the Needles overleaf, you can see a few examples of the selective insertion and removal of information.

3

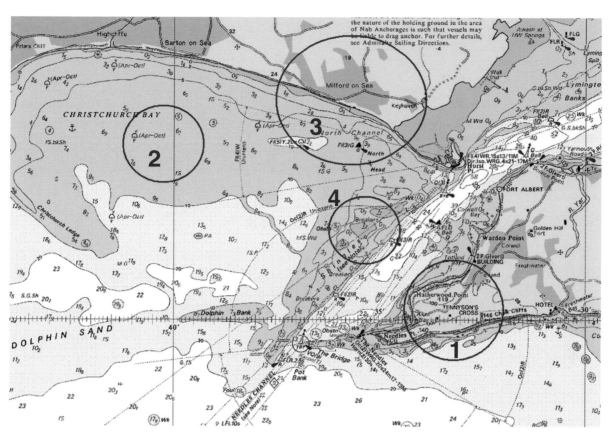

Circle 1

A number of landmarks that will be clearly visible from the sea have been identified: two church towers at Totland, Tennyson's Cross on the cliffs of Tennyson's Down and a clear 119-metre peak above Hatherwood Point. These are all likely to be conspicuous fixing points and landmarks for the mariner.

Circle 2

Many of the soundings that could have been placed in Christchurch Bay have been omitted because the bottom is flat and additional soundings would merely add clutter without providing additional information.

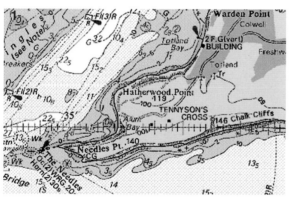

3

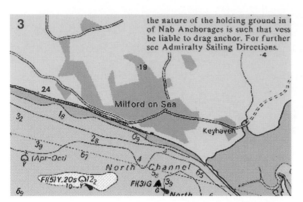

Circle 3

The town of Milford on Sea is depicted only in vague outline to show its limits. Equally, the coast road, which may be seen by a mariner, is depicted; many others have been omitted.

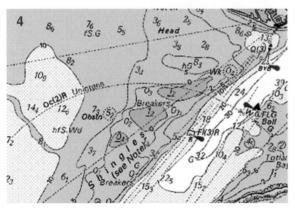

Circle 4

Note the colour coding of the various depth zones over the Shingles, together with the slightly chilling note at the top of the chart: 'The sea breaks violently over the bank in the least swell.'

Even if you fail to read the charted depths or the warning in the margin, the colour shading means that there is little danger that you will mistake this for navigable water.

OVERFALLS

The Needles Channel in the vicinity of The Bridge (50°39'6N 1°36'8W) may be subject to dangerous overfalls in heavy weather at all states of the tide.

NEEDLES CHANNEL
RECOMMENDATION ON NAVIGATION
(50°39'6N 1°36'9W)

The Needles Channel is subject to strong tidal streams and its width is liable to change. Laden tankers over 10 000grt should avoid this channel.

SHINGLES
(50°40'9N 1°36'0W)

Depths over Shingles frequently change and parts of it are occasionally uncovered at MHWS. The sea breaks violently over the bank in the least swell.

4 Other Relevant Documentation

I have a brief but slightly chilling memory of a sailing holiday on the coast of Normandy that my father and I took with a family friend when I was a teenager. We were sailing up to the Iles Chausey from the south on a falling tide one evening in the summer of nineteen sixty-something. The skipper hadn't really done much preparation for this particular landfall. I remember him saying that there was 'bags of tide' and that you can see your way into the sound easily enough. And then we turned towards the shore and headed on in. I'm not certain that he had a large-scale pilotage chart aboard, and I'm sure that he never consulted the *Sailing Directions*, or indeed a local pilot book.

As we got closer, it was touching to see how many people in the anchorage had turned out to wave us in. 'Friendly folk,' we thought. It was only an hour or so later, when we were safely anchored, that someone came over to us in a dinghy and, pointing at a large clump of weed-covered rocks that had been revealed by the falling tide, said how courageous we had been to sail over the rocks. It wasn't so much that people had been waving at us in greeting; more that they were generously trying to help us avoid sailing to certain and uncomfortable doom. We must have cleared the rocks by inches.

There really is some pretty good advice out there in the various publications – particularly relating to the more common landfalls and you have to be pretty pig-headed not to take it. If you overlook this advice, you will put your vessel and its crew at risk. I have no wish to be sanctimonious here: we are all guilty of rushing our preparation from time to time. But if I am honest, many of the more spectacularly embarrassing mistakes that I have made over my seafaring career, both in yachts and in big ships, can be ascribed to a kind of impetuous arrogance that led me to ignore the advice available in the *Sailing Directions* and other supporting publications.

By all means, work out what information you want, and how much risk you are prepared to take. But if you are making a passage in a big ship, you should be aware of the law as specified by the International Convention for the Safety of Life at Sea (SOLAS) 1974[1], which says that:

> All UK registered vessels (including hovercraft and pleasure vessels of over 150 gross tons)... must carry 'Nautical charts and publications to plan and display the vessel's route for the intended voyage and to plot and monitor positions throughout the voyage; an electronic chart display and information system (ECDIS) is also accepted.'

If you've got the supporting publications on board, you might as well refer to them. Small yachts, by contrast, can get by with very much less: on my 12-metre sailing boat, I carry a variety of charts, paper and digital, tide tables and as many good pilot books and sailing directions as I can get hold of. Their information is invariably complementary and they are often quite a good read. In fact, I would almost go so far as to say that parts of the Admiralty *Sailing Directions* are good enough for

---

[1] The 1974 SOLAS Convention has been amended many times over the years. If you need to refer to this document, make sure that you find the most recently amended text.

4

bedtime reading.[2] On a randomly selected page from Chapter One of a now long out-of-date copy of the *India Pilot*, for instance, the following quote describes the inhabitants of the Andaman Islands:

> The aboriginals live in the forest by hunting and fishing . . . and their civilisation is about that of the stone age . . . These tribes are hostile and avoid contact with civilisation. The Jarawas kill on sight.

And further on, the document speaks about the wildlife of India:

> India is the home of about 500 species of mammal, 3,000 species of bird and 30,000 kinds of insect and a wide variety of fish, amphibians and reptiles. An average of about 40,000 people each year are killed by snakes and wild animals.

That's some run ashore: I bet you never expected to find that sort of thing in the *Sailing Directions*. I am distressed to find, however, that the editor of *The Mariner's Handbook*, another invaluable companion of the long night watches, has seen fit in recent editions to omit the stern warning that it carried until a few years back that if you eat too many polar bear livers, you will certainly die through an excess of Vitamin E. (If angry polar bears or environmentalists don't get you first.)

So, what sources of information are available to you? I have compiled the following list principally from the publications available from UK sources, but many other countries have their own equivalents which are every bit as well-constructed and reliable. Indeed, it's striking how similar the charts, the conventions and the supplementary information available to the mariner are, when sourced from competent hydrographic offices around the world.

Catalogue of ADMIRALTY Charts and Publications (NP131)

This publication is invaluable for the serious mariner, but the leisure sailor can probably do without it. It lists all charts and publications available from the UK Hydrographic Office (UKHO) with scale, date of publication and price. The catalogue is revised in December every year. Smaller versions are available in chandleries for local cruising around the United Kingdom. These would be of more benefit to the casual sailor. NP131 is supplemented by the Admiralty Digital catalogue, which carries a full list of charts and services online.

Sailing Directions

The UKHO publishes 75 volumes of the *Admiralty Sailing Directions* in both paper and digital formats, covering the whole world. Loaded with interesting and not-so-interesting information, and written with an almost unbelievable attention to detail, they are designed mainly, but not exclusively, for the big ship driver.[3] But they are always good for a quick browse. *Sailing Directions* are kept up to date through *Notices to Mariners*, and you should always check that you have

---

[2] If, as I suspect, you are thinking that this is the saddest thing that you have read for years, I am in good company. W. Somerset Maugham wrote of the *Sailing Directions* in *The Vessel of Wrath* (Penguin Books, 1951): 'These business-like books take you upon enchanted journeys of the spirit . . . the stern sense of the practical that informs every line cannot dim the poetry that, like the spice-laden breeze that assails your senses when you approach some of those magic islands of the Eastern seas, blows with such sweet fragrance through the printed page.'

[3] The UKHO writes these books primarily for vessels over 12 metres in length.

the latest changes before using the Directions or a supplement. New editions are produced at approximately three-year intervals for most of the volumes. Smaller vessels, in the United Kingdom and elsewhere, should refer to any one of the high-quality pilot and sailing guides on the market, although these volumes are usually not so rigorously updated. If you are using one of these books, check how recently the book has been published: much of this sort of information dates quite rapidly.

The Mariner's Handbook (NP100)

This is my Booker Prize Winner from this list of publications. It's a cracking good read for anyone with a dose of salt in their veins, containing everything from how to correct your charts to characteristics of icebergs, national flags, the COLREGs and pictures of the sea conditions you can expect with various wind strengths. It has much more besides. Available in paper or digital formats, it is updated through *Weekly Notice To Mariners*, and new editions are published every 3 - 5 years. Big

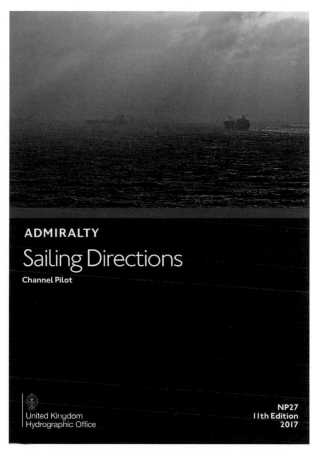

ships should undoubtedly carry this book, and small boat skippers may consider taking it to sea just to have access to the extraordinary range of maritime knowledge that it contains. There are some wonderful vignettes to pass the weary hours of a quiet night watch.

Admiralty List of Lights and Fog Signals

This is very much big ship stuff for the serious mariner: I wouldn't carry a copy in a small leisure vessel. It gives the full details of navigation lights and fog signals anywhere in the world, from light houses to light ships, and buoys whose lights are more than 8 metres above the water. It is updated through *Notices to Mariners* (Section V) and republished annually. For minor changes to lights, *Admiralty List of Lights* will be updated in preference to the chart – so this always carries the authoritative data.

Admiralty List of Radio Signals

Another set of volumes that are not for the faint-hearted. It, too, is designed for the professional mariner. There are six volumes, containing a wealth of useful information for the long-distance mariner: coast radio stations (including medical advice by radio), radio navigation aids and time signals, safety information services, GMDSS, port service radio frequencies and maritime communications frequencies.

4

Tide Tables

Everyone should have one of these, either in the original version of the *Admiralty Tide Tables*, or in a facsimile extract like *Reeds Nautical Almanac*, which is designed for small boats and contains secondary port information.[4] Despite some pretty ferocious yachtsman's folklore about secondary ports, it is not at all difficult or time-consuming to work out the height of tide in any port – and the ability to accurately calculate height of tide is an indispensable competence if you want to stay afloat. The data contained in the *Tide Tables* for the Standard Ports is generally derived from observations over an extended period, normally more than one year's worth of continuous data. Secondary ports often have very much sketchier information, and you should consider this data less reliable. The *Tide Tables* contain a number of other pieces of useful information, including the data that allows you to calculate heights of tide from first principles, either by hand or by computer. If you haven't tried this, and you have a bit of time on your hands, get hold of a copy of the *Tide Tables* and have a go: it isn't particularly complicated, and it can be highly satisfying.

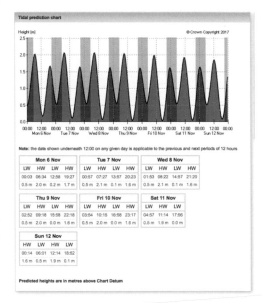

For those with slightly less time available, and particularly for leisure mariners, the UKHO 'EasyTide' app is a really valuable way of deriving accurate height of tide predictions online, for a large number of coastal locations around the world. It provides free predictions for the next 7 days; longer-term predictions are available on subscription. This is a prediction that I downloaded for Ushuaia in Argentine Patagonia:

Tidal Stream Atlas

There are 22 *Tidal Stream Atlases* published by the UKHO covering UK waters and NW Europe. They provide a valuable visual representation of the flow of water, although they do little more than repackage information that is available through the tidal diamonds on most large-scale charts. Each atlas contains an hour-by-hour diagrammatic representation of tidal flow, referenced to the time of high tide at a nearby Standard Port. They are undoubtedly useful to have on the bridge or in the cockpit of any vessel.

Nautical Almanac

The term 'almanac' covers a variety of publications. At the more high-brow end of the market, the UKHO *Nautical Almanac* contains all the data that you need in any particular year to conduct astronavigation at sea. An alternative, and rightly very popular, source of general information for smaller vessels in UK waters and around NW Europe is the excellent *Reeds Nautical Almanac*. This does not give astronavigation tables, but contains virtually everything that you need for local

---

[4] *Reeds Nautical Almanac* is carried onboard all RNLI lifeboats operating in the United Kingdom and Republic of Ireland.

4

pilotage, navigation, meteorology, communications and quite a lot more. It also contains tidal data for Standard and Secondary Ports in the area. There are a number of first-class books in a similar vein published in other countries.

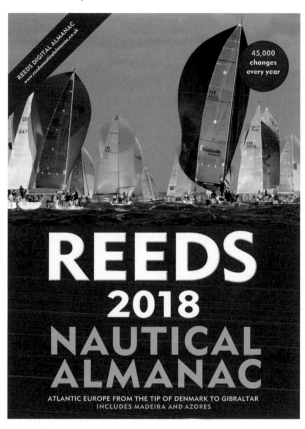

Local Knowledge

Almost every port authority around the world will insist that large ships are brought into and out of harbour under the eye of a local pilot. I have experienced many pilots over the years; the majority of them good, but a sizeable minority vary from mediocre to downright dangerous. Local knowledge, when provided by an informed, competent and experienced mariner, is utterly invaluable – there are few places where the tidal streams don't have the odd quirk, the weather is utterly predictable and the tugs and riggers always impeccably behaved. But they are not a substitute for meticulous preparation. I remember taking a large naval vessel into a north European port, with a long and seemingly perilous transit through a rock-strewn estuary . . . and the only time the pilot opened his mouth was to take a sip of his coffee. I have to assume that he was struck dumb by the standard of my navigation.

Notices to Mariners

Professional mariners update their charts regularly; amateurs seldom seem to. I remember going alongside a French yacht lying alongside us in the Morbihan in southern Brittany. As soon as we were tucked up, the skipper had asked me to come over and show him where the navigation lights were situated. It was a strange request, especially coming from a local, and I went across with my chart, at that stage only a few months old. When I arrived, I discovered that his most recent chart of the area was a black-and-white fathoms chart dating back to 1958. Needless to say, the lay-down of navigation lights had changed dramatically since it had been printed. I asked him whether he felt just a little vulnerable with this chart (and doubtless many more just like it) and he shrugged and said that they got by.

You only need one document to help you keep your charts up to date and that is the *Weekly Notices to Mariners*, available free to all users of Admiralty charts and books, and available online from the UKHO.

I will talk about how to keep a chart updated in Chapter 5.

ADMIRALTY NOTICES TO MARINERS

Notices 4838–4944/17

United Kingdom Hydrographic Office

Weekly Edition 42
19 October 2017
(Published on the ADMIRALTY website 9 October 2017)

CONTENTS

I	Explanatory Notes. Publications List
II	ADMIRALTY Notices to Mariners. Updates to Standard Nautical Charts
III	Reprints of NAVAREA I Navigational Warnings
IV	Updates to ADMIRALTY Sailing Directions
V	Updates to ADMIRALTY List of Lights and Fog Signals
VI	Updates to ADMIRALTY List of Radio Signals
VII	Updates to Miscellaneous ADMIRALTY Nautical Publications
VIII	Updates to ADMIRALTY Digital Services

For information on how to update your ADMIRALTY products using ADMIRALTY Notices to Mariners, please refer to NP294 How to Keep Your ADMIRALTY Products Up-to-Date.

Mariners are requested to inform the UKHO immediately of the discovery of new or suspected dangers to navigation, observed changes to navigational aids and of shortcomings in both paper and digital ADMIRALTY Charts or Publications.

The H-Note App helps you to send H-Notes to the UKHO, using your device's camera, GPS and email. It is available for free download on Google Play and on the App Store.

The Hydrographic Note Form (H102) should be used to forward this information and to report any ENC display issues.

H102A should be used for reporting changes to Port Information.

H102B should be used for reporting GPS/Chart Datum observations.

Copies of these forms can be found at the back of this bulletin and on the UKHO website.

The following communication facilities are available:

NMs on ADMIRALTY website:	Web:	admiralty.co.uk/msi
Searchable Notices to Mariners:	Web:	www.ukho.gov.uk/nmwebsearch
Urgent navigational information:	e-mail:	navwarnings@btconnect.com
	Phone:	+44(0)1823 353448
	Fax:	+44(0)1823 322352
H102 forms	e-mail:	sdr@ukho.gov.uk
(see back pages of this Weekly Edition)	Post:	UKHO, Admiralty Way, Taunton, Somerset, TA1 2DN, UK
All other enquiries/information	e-mail:	customerservices@ukho.gov.uk
	Phone:	+44(0)1823 484444 (24/7)

For UKHO use only 217442

Contents of the Weekly Notices to Mariners

The *Weekly Notices to Mariners* is split into eight sections:

Section I. Weekly Notes. Publications List.

Section II. Admiralty Notices to Mariners. Updates to Standard Nautical Charts.

Section III. Reprints of NAVAREA I Navigation Warnings.

Section IV. Updates to Admiralty Sailing Directions.

Section V. Updates to Admiralty Lists of Lights and Fog Signals.

Section VI. Updates to Admiralty Lists of Radio Signals.

Section VII. Updates to Miscellaneous Admiralty Nautical Publications.

Section VIII. Updates to Admiralty Digital products and services.

Cumulative List of Notices to Mariners

This is a remarkable document that records the edition date and the change state of every single UKHO nautical chart. If you want to check that your charts are up-to-date (and you probably do, from time to time), you have a number of choices:

- You can use the QR code at the bottom left hand margin of the chart.

- You can look on the UKHO website at https://www.ukho.gov.uk/nmwebsearch/

- Or you can check out the chart correction state from the *Cumulative List of Notices to Mariners*.

The *Cumulative List* is published every six months, and with increasing digitisation it is probably becoming less useful to mariners.

Annual Summary of Notices to Mariners

The big brother of *Weekly Notices to Mariners* and the *Cumulative List* is the *Annual Summary of Notices to Mariners*, published in January of each year and available online. It too contains a useful range of information, and I would recommend that just about any mariner, professional or amateur, gets hold of a copy. At the very least, browse through the contents online. It also contains details of changes to the *Sailing Directions*, and any *Temporary and Preliminary Notices to Mariners* (see Chapter 5) that are still in force at the start of the year.

Port Approach Guides

Recognising that port entry and exit is the time of both highest risk and highest workload for the bridge team in any ship, the Hydrographic Office has produced a range of 800 *Port Approach Guides*, printed on full-sized chart paper. These valuable documents display with admirable clarity the essential information that you need to navigate through the approaches to some of the world's busiest ports. Using QR codes and signal flags to indicate specific warnings, they give you the information that you need to know in a simple, clear and visual format. They are well worth looking at if you are responsible for navigating large ships in and out of busy ports.

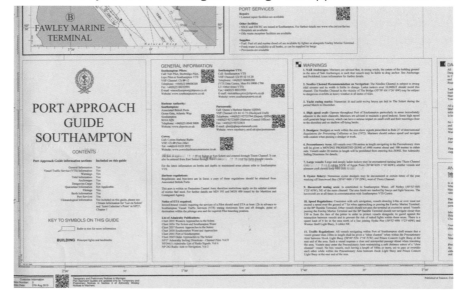

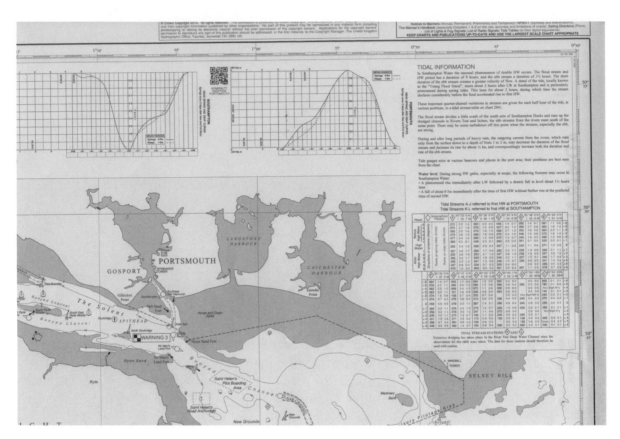

Spoilt for Choice

There is quite a lot of competition in the marketplace – for digital and paper charts, and for the supporting publications too. I am biased; I have been brought up on Admiralty publications from the start, and I would not move away – although I happily supplement these with high quality publications from other sources that I have grown to trust over the years. You must make your own choices, but it is important to satisfy yourself, in advance, that the integrity of the information that you are buying is as good as you can get for your purposes, and that the update service is both timely and comprehensive. Very often, you get what you pay for, and cheap information may not always prove to be an economy in the long run.

5 How to Use a Chart

If you are going to pay quite a lot of money to buy a chart, it is worth keeping it in a good state so that you can use it again and again. Charts are printed on high-quality paper and there is no reason why they shouldn't last for many years.

First of all, when storing a chart:

- keep it dry

- keep it flat

- fold it down the existing folds to reduce distortion.

When navigating on a chart:

- Use a 2B pencil, or softer.

- Always rub the chart out after use using a good-quality soft rubber.

- Never – and I mean never – place a coffee cup on the chart. Or a wine glass for that matter. You run the risk of permanently staining the chart, which could obscure vital information.

- If it is unavoidable that your chart will get wet, try putting it in a large plastic envelope (such as the UK Hydrographic Office provides for the Small Craft Folios) and navigating with a wax pencil, but take care because your navigation is likely to be less accurate this way – particularly if the chart slips around inside the envelope.

In General

Keep the chart up to date and change it when a new edition is published. Read any supporting documents: pilots, etc. so that you can put the chart into context. In particular, give yourself time to properly study a chart before you use it: read the notes, look at the projection, check the consistency of soundings and note the bottom topography.

A really simple mnemonic about chart scales:

SMALL SCALE – LARGE AREA

LARGE SCALE – SMALL AREA

Visualising the Chart

A chart is merely a two-dimensional representation of three-dimensional space, so you need to take time to visualise the layout of any pilotage waters in advance, in order to recognise the features and landmarks more easily while you are under the inevitable pressure that comes with inshore

5

pilotage. In my experience, it doesn't matter how thoroughly you prepare for pilotage, you will always be surprised by something!

One way to visualise the entry to a port is to look through the *Sailing Directions* and pilot books for aerial photos of ports and stretches of coastline. These are invaluable when approaching an unfamiliar port as a way of working out the relationship of various features, and visualising how they will appear from the bridge.

Cowes, Isle of Wight: aerial photos are really useful when used alongside a chart (© Sealand Aerial Photography Ltd / Fernhurst Books Limited)

Some charts carry small line sketches of the coast from seaward. These are usually engravings which were produced some time ago, and they are now reproduced only when they are both accurate and relevant. An example from a chart of the Isles of Scilly is shown on the following page.

And finally, you should never overlook the humble seaside postcard. Bizarre as it may seem, local postcards often carry remarkably useful views of the coast, estuaries and harbours (usually taken from the air) and often at a variety of tidal states. In my sailing boat I now collect these and paste them into the *Sailing Directions* to supplement the published pictures. This is a postcard that I bought in Audierne in South Brittany, which illustrates the whereabouts of navigable water and the sandbar, and which served as a useful guide in my yacht when entering harbour. (The banks may have moved over the intervening years - so please don't use it for navigation!)

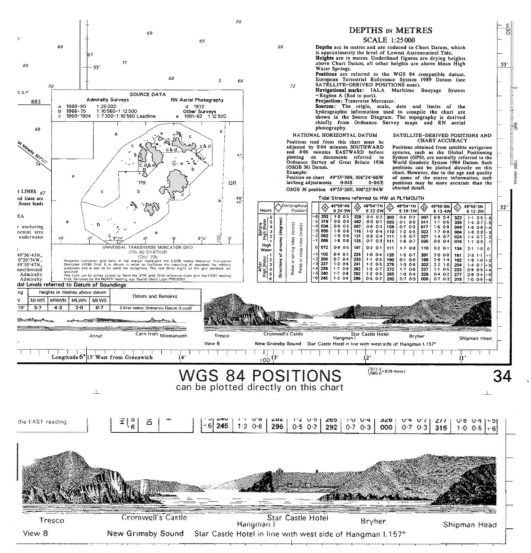

A difficult approach to the Isles of Scilly is assisted by these engravings

In any case, before you find yourself in a tight, unfamiliar and often busy waterway, you should take the time to understand:

■ what the various marks look like

■ how the tidal stream is likely to be setting

■ how much water you will have under the keel

■ where you would be unsafe to venture

■ what the principal traffic flows are

■ restrictions or limits to navigation

■ and how, in broad terms, it all fits together.

With this resolved, you are likely to be in a position to react more appropriately when something unexpected occurs during the pilotage itself. It will be time well spent.

5

Chart 5011: Symbols and Abbreviations Used on Admiralty Paper Charts

The information from *Chart 5011*, the Admiralty Chart Symbol reference publication, is reproduced at the end of this book (page 127). Every vessel using a chart should carry a copy. Personally, I have always considered it indispensable because I have never found it easy to remember the precise meaning of all the symbols, and it helps me get that little bit more from my charts. You will find that there are differences between the symbols used on charts in the international folio and those that are published solely in domestic folios. There is no need to get too distracted by this: all of the symbols from both chart sets are pretty intuitive, and you are unlikely to get confused.

Chart Corrections

Updating charts is, I think, a bit like knitting: it is time-consuming, it's repetitive and for much of the time you feel that you would be better employed doing something else. But without wishing to put too fine a point on it, chart correction really isn't a discretionary activity, so we all need to find time to fit it into our schedules – weekly for commercial vessels and at appropriate intervals for yachtsmen.

The UKHO receives thousands of change notifications every week from a huge variety of agencies all over the world – far more than it could realistically send out to its users without completely swamping them. So whenever the UKHO is notified of a correction anywhere in the world, it rapidly takes steps to check the accuracy, the relevance and the urgency of this information.

Once it has established that the data is correct, it separates the safety-critical changes from the rest. Changes that are not safety-critical are held until the next edition of the chart is published. All safety-critical changes, however, are sent out in the *Weekly Notices to Mariners*. The most urgent of these will also be distributed as Navigation Warnings and materialise on our NAVTEX machines, well in advance of their subsequent publication in *Weekly Notices to Mariners*.

The UKHO's criteria for classifying changes as 'safety-critical':

1. New dangers which are significant to surface navigation (e.g. shoals and obstructions with less than 31 metres of water over them, and wrecks with a depth of 28 metres or less).

2. Changes in general charted depths that are significant to submarines, fishing vessels and other commercial operations.

3. Significant changes to the critical characteristics (character, period, colour of a light, or range) of important aids to navigation.

4. Changes to, or introduction of, routing measures

5. Works in progress.

6. Changes in regulated areas, for example restricted areas or anchorages.

7. Changes in radio aids to navigation.

5

8. Addition or deletion of a conspicuous landmark.

9. In harbour areas: changes to wharves, reclaimed areas, updated date of dredging, works in progress, new ports or port developments.

10. In UK home waters, all cables or pipelines both overhead and seabed, to a depth of 200 metres. Outside UK home waters, all cables or pipelines, seabed telecommunications cables to a depth of 40 metres, seabed power cables and pipelines to a depth of 200 metres.

11. Offshore structures, for example production platforms, wind turbines, marine farms.

12. Pilotage services.

13. Vertical clearances of bridges. Also horizontal clearances in US waters.

Temporary and Preliminary Notices to Mariners

The *Permanent Notices to Mariners* report establishes changes that are likely to be enduring. In addition, the UKHO publishes *Temporary and Preliminary Notices* (Ts and Ps). These are exactly what they say on the tin: Temporary Notices are used when information will remain valid for only a short period. Preliminary Notices are used when early promulgation of a change is needed to alert mariners that:

- Action or work will shortly be taking place.

- Information has been received that is too complex to promulgate through *Notices to Mariners* (NTMs): the *Preliminary Notice* will give a précis of the information, and advance warning of a new chart or edition.

- Further confirmation of details is necessary before an NTM is issued.

- Ongoing work is under way (for instance the construction of a bridge) that will be published as an NTM when work is complete.

All 'Ts and Ps' that are in force on 1 January each year will be reprinted in the *Annual Summary of Notices to Mariners*.

Weekly Editions of NTMs

Weekly Notices to Mariners are provided free of charge to users of Admiralty charts, and can either be received in paper from or downloaded from the UKHO website (www.admiralty.co.uk). I would strongly recommend using the QR code in the bottom left-hand corner of any Admiralty chart to check that it is up-to-date before you start using it.

When you buy a new chart from an authorised agent, it should be accurate and up to date for all *Permanent Notices to Mariners*, but not for the Ts and Ps. In any case, you should always check. There are 3 ways of checking if a paper chart has been updated:

- From the UKHO website.

- Using the QR code on the chart.

- Or looking up the chart in the *Cumulative Notices to Mariners.*

Paper and digital charts supplied from other publishers will have their own bespoke arrangements. It's best to check online, or with the retailer, to find out the details.

What a Notice to Mariners Looks Like

Each Notice has a unique number (in this case 4702). These numbers reset to zero at the beginning of each year, so technically as a 2017 Notice it will be 4702/17, to differentiate from 4702 in other years.

The title carries a short descriptor, which identifies the location and the nature of the change. This is followed by information on the source of the data and any relevant notes (ENGLAND – South Coast – Portchester Lake South and Spit Sand – Depths).

The main block of the Notice starts with the Admiralty and International Chart

II

4702°   **ENGLAND - South Coast - Portchester Lake S and Solit Sand - Depths**
Source: British Government Survey

Chart 2036 (INT 1730) *[previous update 2794/17]* **ETRS89 DATUM**
Replace   depth *1*, with depth *0*   50° 46° 58N, 1° 06° 21W

Chart 2037 (INT 1731) *[previous update 3794/17]* **ETRS89 DATUM**
Replace   depth *1*, with depth *0*   50° 46° 58N, 1° 06° 21W

Chart 2045 *[previous update 4033/17]* **ETRS89 DATUM**
Replace   depth *1*, with depth *0*   50° 46° 58N, 1° 06° 21W

Chart 2625 *[previous update 3794/17]* **ETRS89 DATUM**
Insert   depth *0*
Delete   depth *1*, close SE of: (a)   50° 46° 58N, 1° 06° 207W
                                                                    (a) above

Chart 2629 *[previous update 3794/17]* **ETRS89 DATUM**
Insert   depth *8*
Delete   depth *9*, close NW of: (a)   50° 48° 685N, 1° 07° 125W
                                                                    (a) above

Chart 2631 (INT 1732) *[previous update 3794/17]* **ETRS89 DATUM**
Insert   depth *8*
Delete   depth *9*, close NW of: (a)   50° 48° 685N, 1° 06° 125W
                                                                    (a) above

Catalogue numbers (Chart 2036 (INT 1730)), before moving on to the precise details of the change. You will see that this correction affects a number of charts. With each chart, the number of the previous change has been attached (in this case 3794/17) so that you can check you are not missing a vital change. It also tells you that the change is made in accordance with ETRS89 datum, not the more common WGS84. This is handy, because the European Terrestrial Reference System 1979 (ETRS79 to you and me) is indistinguishable from WGS84.

Changes come with any one of five instructions: 'Insert', 'Amend', 'Substitute', 'Move' and 'Delete'. Only the changes given in Section II of the *Notices to Mariners* (Updates to Standard Nautical Charts, i.e. the Permanent Notices to Mariners) should be inserted in ink. Everything else, including Temporary and Preliminary Notices, Navigation Warnings and Notices from other sources should all be inserted in pencil until verified formally in the Weekly NTMs. When they are no longer relevant, make sure that you rub them out.

Block Corrections

Hand-drawn changes are fine, but most of us have a pretty low tolerance to complicated amendments and, of course, the more complex the change, the more likely one is to make a mistake in plotting it. To get over this problem, when adding or removing significant amounts of information, the *Weekly Notices to Mariners* publishes those changes as 'block corrections': small chartlets which are designed to be cut out and pasted onto the appropriate part of the chart as an overlay. If you download the NTMs online, you will need to print out relevant block corrections (which are promulgated as .pdf files) on good-quality paper with a waterproof ink, if possible.

Updating Other Chart Series

1. **Admiralty Leisure Series**. This series, which comes as single charts or in A2 folios, is updated weekly, using Leisure notices that can be downloaded from www.ukho.gov.uk/leisure

2. **Imray Charts** and book supplements may be updated using the website at http://www.imray.com/corrections.cfm

3. **Corrections to NOAA and Canadian Charts** are available online at http://www.nauticalcharts.noaa.gov

And What Happens if You Want to Submit an Update?

If you find something which isn't charted and should be, or if you have information on a port that isn't listed in the *Sailing Directions*, the UKHO encourages you to submit a Hydrographic Note, the instructions for which, as well as the form, can be found towards the end of each edition of the *Weekly Notices to Mariners*. You can also find the form in *The Mariner's Handbook*, or download it from the UKHO website. As if that was not enough, there is even a handy Admiralty H-Note App which you can use with Android and Apple devices.

Over and above the need to report navigational anomalies, SOLAS requires the master of every ship to report any of the following:[1]

■ dangerous ice

■ a dangerous derelict

■ any other danger to navigation, including shoal soundings, uncharted dangers and navigational aids out of order

■ a tropical storm or winds of Force 10 and above, for which there is no warning

■ air temperature below freezing associated with gale force winds causing severe icing.

These reports should be made to ships in the vicinity and to the nearest coast radio station.

---

[1]*See The Mariner's Handbook*, Para 3.1.

6 Orientation

The difficulty with producing charts is that, at the end of the day, you are having to create an accurate representation of a curved object (the surface of the earth) on a flat surface (the chart paper). This inevitably results in compromise, but the solution which is generally accepted to hold the least compromise for the everyday mariner is the Mercator projection. The vast majority of the charts that you will use at sea will be plotted using the Mercator projection, or one of its derivatives.

You will always find the type of projection that is being used in the Title Block of a chart.

Mercator Projection

It is perhaps surprising that the Mercator projection, which was developed in 1569 by Gerardus Mercator, is still in use today, when many other aspects of maritime cartography have changed so radically.

If you can imagine it, the Mercator projection wraps a very large piece of paper around the equator, leaving the ends, which are aligned north–south, open. It then puts an immensely strong light in the centre of the earth and projects the shadows of the land mass onto the paper.

MERCATOR PROJECTION

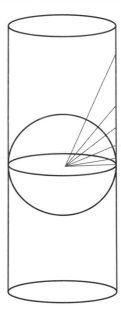

As you can imagine, there is very little distortion of the image close to the equator, where the paper touches the surface of the earth, but the further north or south you go, the more the latitude scale is stretched. The poles don't get on the paper at all.

The Mercator projection has a number of crucial advantages on any other projection:

■ The lines of latitude and longitude always cross at right angles.

■ A rhumb line, the straight line between two points on a chart, cuts all the lines of longitude at a constant angle, making short-range navigation very much simpler.[1]

■ Angles on the earth's surface are the same as the angles on the chart.

■ Lines of longitude are equally spaced across the whole chart, although lines of latitude are given a variable spacing: wider as you move away from the equator.

However, there is one very great disadvantage to the Mercator projection, too: the distance scale, which is read directly off the latitude scale, varies with latitude. Check it out on the next Mercator chart that you come across: in the Northern Hemisphere, 20 miles (or 20 minutes of latitude) on your dividers at the top of the chart will span a greater distance on the chart than 20 miles at

---

[1] Over long distances, a great circle route will provide a shorter distance between two points: on a Mercator projection, this appears as a curved line.

the bottom of the chart (the opposite applies in the Southern Hemisphere). **So you have to be absolutely rigorous when using a Mercator chart to measure distance against the latitude scale adjacent to the part of the chart that you are using.**

The Mercator projection accounts for the great majority of charts in everyday use today. It does, however, have a valuable younger brother, the Transverse Mercator Projection, which is a clever device used to reduce distortion on large-scale (small area) charts.

Transverse Mercator Projection

This is quite ingenious. You will recall that the Mercator projection has very little distortion close to the equator, where the cylindrical sheet of paper, so to speak, touches the earth's surface.

Suppose you are interested in making a chart of an area which is some way from the equator. Why not rotate the cylinder by 90 degrees and set it up so that it still touches the earth's surface, but along the line of longitude that passes through your chosen point and the two poles?

You will get very little distortion anywhere along this line of longitude, irrespective of latitude.

You have, in effect, created a chart with the advantages of the Mercator projection, but without the 'latitude stretching effect' away from the equator and, as long as you stay close to your chosen meridian you will have very little distortion over and around your chosen area.

TRANSVERSE MERCATOR PROJECTION

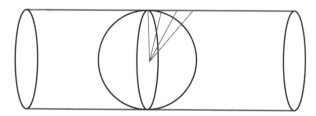

This projection would be impractical for small-scale charts (large area) because the distortion of longitude over greater distances would make it quite confusing for the mariner. It is, however, widely used for large-scale charts of bays, anchorages and harbours, and as a large-scale chart it is, from the user's perspective, indistinguishable from the Mercator projection.

Gnomonic Charts

Gnomonic charts (pronounced 'no-mon-ic') are not common, but you do see them around from time to time. They are formed by taking a flat plotting sheet and allowing it to touch the earth's surface at a single point (normally the point that you are interested in charting). You then put a very strong light at the centre of the earth and plot out the shadow of the earth's surface on the paper. This produces a chart on which all of the meridians of longitude converge on the pole as straight lines, with the lines of latitude appearing as arcs of a circle, cutting the lines of longitude at right angles.

GNOMONIC PROJECTION

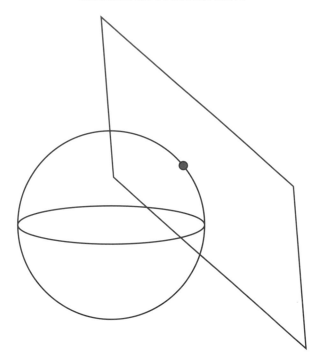

Gnomonic charts are normally used for charting small areas, close to the point of contact between the earth and the plotting sheet, where distortion is minimal.

Plotted on a small scale, where they cover large areas of ocean, they have two specific uses. First, they are a helpful way of plotting the polar regions, and they are used by those navigating in high latitudes. Second, and of more practical importance to the majority of us, they have the property of displaying Great Circle routes as straight lines. So if you have a chart of the Atlantic and want to plot the shortest distance between, say, Plymouth and Boston, you only need a pencil and a ruler. You draw the line on a gnomonic chart and transfer the intermediate points from the gnomonic chart to your navigation (Mercator) chart.

You may encounter other projections from time to time, but these three are the most common and, of these three, the straightforward Mercator projection is by far the most widely-used and the most useful.

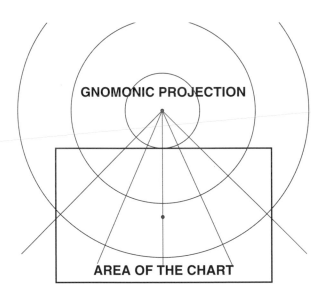

GNOMONIC PROJECTION

AREA OF THE CHART

Latitude, Longitude and Distance

Latitude scales are displayed along both sides of the chart, and longitude scales along the bottom. As we have already established, the Mercator projection gives a variable latitude scale, which changes as you move from the bottom of the chart to the top. The longitude scale is constant across the whole spread of a Mercator chart.

Writing Latitude and Longitude

A degree of latitude or longitude is divided into 60 minutes, and each minute into 60 seconds. The symbol of a degree is °; for a minute is '; and for a second is ". It is, however, increasingly common to ignore seconds and express position as degrees, minutes and decimal points of a minute, followed by the letters 'E', 'W', 'N' or 'S'. So, the position of Longstone Lighthouse off the coast of

Northumberland is written: **55° 38'.62N 001° 36'.66W**.[2] Note that the minute or second symbol is, by convention, always put after the whole number of minutes and seconds, not after the decimals.

The important thing to recognise is that, anywhere in the world, a minute of latitude is exactly equal to one nautical mile or approximately 6 076 ft (1 852 metres).[3] A tenth of a mile, which is about 200 yards, is generally known as a 'cable'.

On a more practical note, I would always recommend that you check and double-check the spacing between the graduations of latitude and longitude on any chart you use, because the layout and the scale may well change from chart to chart. The 2 charts below both cover the Shetland Isles, but they are of very differing scales, and this is reflected in the way that the latitude and longitude scales are drawn in the margins. Chart A is a large scale chart, with both latitude and longitude graduated to the nearest tenth of a minute. Chart B is a smaller scale chart of the same general area, with graduations to half a minute. It is also worth noting that, as Shetland lies close to the Greenwich Meridian, the polarity of the longitude scales may reverse as you move from one chart to the other.

6

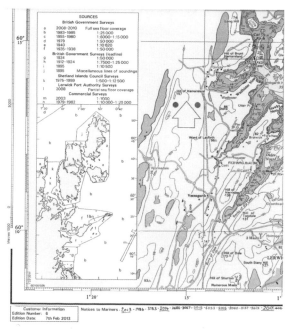

Chart A

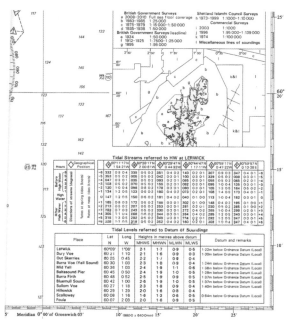

Chart B

Angular Measurement

Angles on a Mercator chart will be exactly the same as the equivalent angles on the earth's surface, and all charts which originate in the UK Hydrographic Office are aligned to true north. You will find a number of compass roses scattered helpfully around the chart to allow you to measure angles and bearings.

---

[2]Alternatively, but very much less commonly, it can be written: **55° 38' 37".2N 001° 36' 39".6W**.

[3]You may have learnt, as I did, that a nautical mile is 6 080 ft. Because the earth is not a perfect sphere, the length of a minute of arc varies slightly between the pole and the equator. The difference is not significant to the likes of you and me.

6

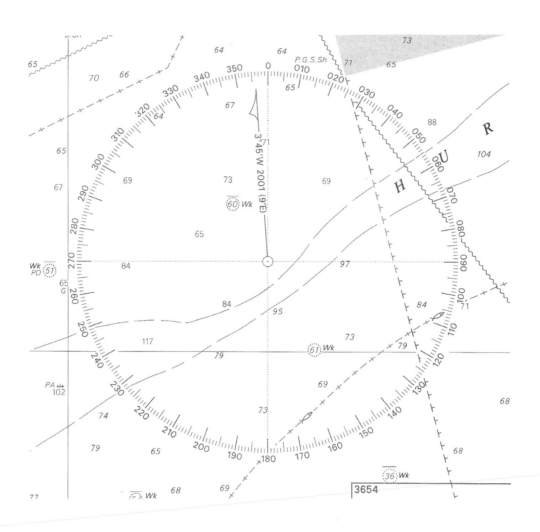

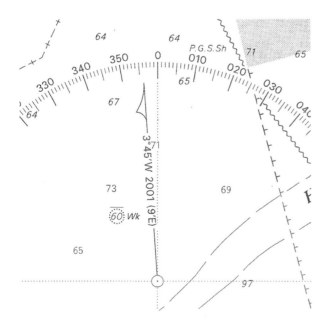

'Variation', the difference between true and magnetic north, changes from year to year. It is generally shown in the compass rose by a subordinate arrow pointing at magnetic north, and a factor to allow the annual changes in variation to be calculated.

In this case, the variation is shown as **3° 45'W 2001 (9'E)**. That is to say that the variation was **3° 45'W** on 1 January 2001, moving **9'** east every year. So, by 1 January 2010, it has moved **81'** east (or **1° 21'**) and the variation had become **2° 24'W**.

6

On smaller-scale charts you will find magenta lines of equal variation drawn as a sort of curved diagonal across the chart, normally where the variation is a whole number of degrees east or west. It is easy enough to interpolate between them if you need to. These lines of equal variation are known as **'isogonals'**.

This chart extract shows a series of isogonals overlaid on an oceanographic chart (Chart 4713) with a variation that would undoubtedly be of interest if you were navigating by magnetic compass.

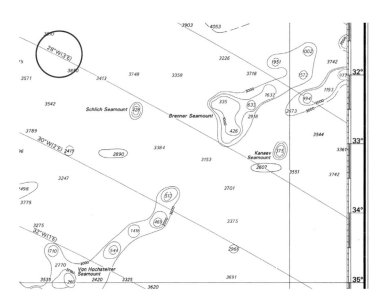

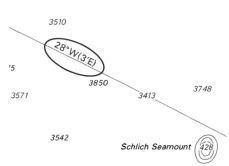

Local Magnetic Anomalies

Finally, you may occasionally come across local magnetic anomalies which have a perimeter that can only be described as 'wobbly'. The value inside is the greatest divergence that you can expect on a magnetic compass as you transit that area.

LOCAL MAGNETIC ANOMALY

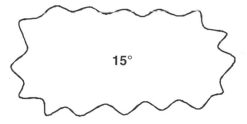

7 The Basics (and Where to Find Them)

Take any Admiralty chart, unfold it on the table and you already know quite a lot about it. Even with a cursory glance, you will notice the title, which informs you of the area covered by the chart (the title font is the largest on the chart) and you will be able to get an appreciation for the shape of the land (printed in yellow) and the areas of shoal water (variously in green, dark blue and pale blue).

You will know that 'true north is up', because that is the way that all charts originating in the UK Hydrographic Office (UKHO) are printed, and the very fact that the chart is printed in yellow, green and blue will tell you that the soundings and heights are calibrated in metres. Clearly, a grey, blue and white colour scheme would indicate a chart calibrated in feet and fathoms.

Perversely, the first part of a chart that you should look at is its margins. Marginal information is often missed by mariners, but is actually more important and more interesting than you might think.

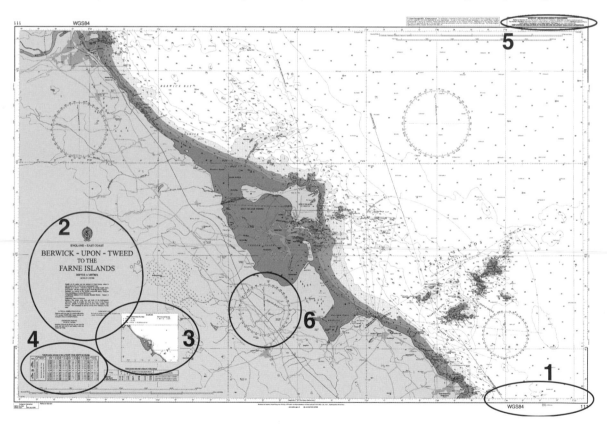

The chart above, for instance, is self-evidently a metric chart, but what else can we find out about it?

7

Circle 1

Circle 1 in the bottom right-hand corner of the chart, tells you that the chart's number is 111[1] and it has been plotted to the WGS84 datum, which means that you don't need to apply corrections when plotting positions derived from the US GPS system. Occasionally, there will be a second chart number alongside the UKHO number; something like 'INT 1234'. This would be the reference number for this chart within the International Chart Series.[2]

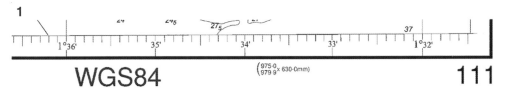

Adjoining charts of the same scale are identified in the appropriate margin of the chart to make it easier for you to run across from one chart to the next, and larger-scale charts of a small, enclosed, area are delineated by a magenta square with the chart number printed inside. In this case, you will find the large-scale harbour plan of the port of Seahouses, in greater detail, on Chart 1612.

Circle 2

The Title Block is the principal source of supplementary information on any chart, and you should always look at this.

In this case, you will see the seal of the UKHO, but you may additionally find the seal of other foreign hydrographic offices which played a part in producing the chart.

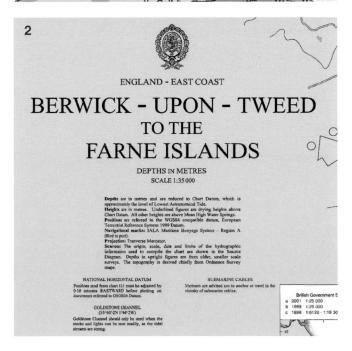

---

[1] Sometimes, usually when working in an international context, British Admiralty charts are given the prefix 'BA', as in 'BA111', to differentiate them from charts produced by other agencies carrying the same number.

[2] The International Chart Series is being developed by the International Hydrographic Office with a view to unifying as many chart systems as possible.

Below the seal is a geographic locator: 'ENGLAND – EAST COAST', followed by the title, which is the chart's main descriptor. The title nearly always describes the coverage of the chart; in this case 'BERWICK-UPON-TWEED TO THE FARNE ISLANDS'. You have already established that this is a metric chart from its colouring, but this is confirmed in the next line, followed by the scale of the chart.

Scale and Projection

It is probably worth paying a little attention to the scale of the chart. This particular chart is drawn to a scale of 1:35 000, which is pretty typical of a coastal passage chart. That is to say, one metre on the chart is equivalent to 35 kilometres (35 000 metres) on the ground.

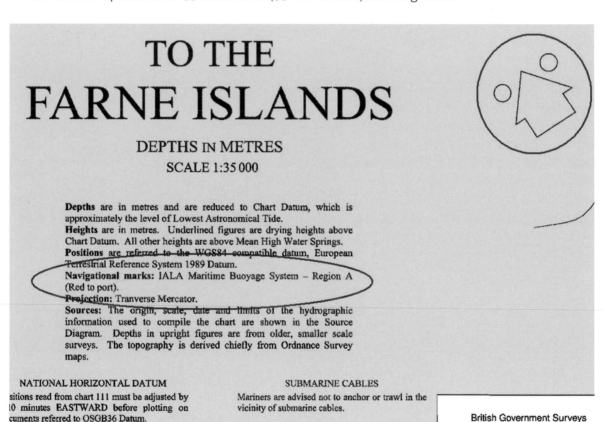

We have already established that distance on a chart is measured using the latitude scale (down the sides of the chart) on the basis that

> one sea mile = 1 minute
> of latitude

or 1/60th of a degree of latitude. And 1/10th of a mile, known as a cable, is about 200 metres.

Mercator charts stretch latitude as you get closer to the poles. That means that you must always measure distance using the latitude scale adjacent to the part of the chart that you are using. Because of this north–south scale variation, Mercator charts always establish a latitude at which the scale applies, e.g.: 'SCALE 1:50 000 at 51° N'.

7

The chart in this illustration, however, has been plotted using a Transverse Mercator projection (see Chapter 6), which has the advantage of giving very little distortion across a large-scale (small area) chart. So the scale can be assumed constant across the whole area of the chart.

You should note the system of buoyage in this area and its orientation, which I will discuss more fully in Chapter 9. The Title Block points out that the system of buoyage in use is IALA Region A (i.e. you should leave the red marks to port when travelling with the flood stream). And the hollow magenta arrow with two circles shows you which way the flood tide is assumed to flow in this area.

The remainder of Circle 2 is pretty straightforward, but I would advise that you read *all* of the individual warnings with care, on the grounds that the hydrographer would not bother printing them if they weren't important. On this chart, for instance, you will find a warning relating to the Goldstone Channel, just to the east of Holy Island, which reads: 'Goldstone Channel should only be used when the marks and lights can be seen readily, as the tidal streams are strong.' You can assume that this has been written for a good reason.

NATIONAL HORIZONTAL DATUM

Positions read from chart 111 must be adjusted by 0·10 minutes EASTWARD before plotting on documents referred to OSGB36 Datum.

GOLDSTONE CHANNEL
(55°40'·2N 1°44'·2W)

Goldstone Channel should only be used when the marks and lights can be seen readily, as the tidal streams are strong.

Circle 3

On Admiralty charts[3], you will always find a Source Data Diagram tucked away somewhere in the margins, telling you when, and how thoroughly, the chart was surveyed. **This is important reading** because, more than anything else, it tells you how much faith you can put in the information that the chart contains. This particular chart appears to be quite thoroughly surveyed to modern standards, with the exception of a small area, marked 'c', to the south of Holy Island which was surveyed in 1898 with a leadline. In that area, you would treat the soundings with greater suspicion, and keep a look out for uncharted dangers.

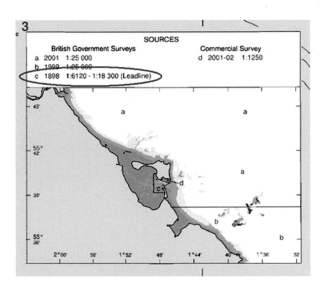

Increasingly, new charts are using ZOC codes to describe the thoroughness and accuracy of the survey data. You will find the table of the ZOC codes on page 124.

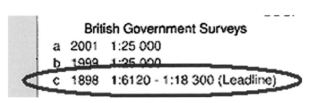

---

[3]You will not always find a source diagram on more commercial charts.

In addition to the Source Data Diagram, there are a number of other clues that indicate a poor quality of depth information on a chart:

- Broken or incomplete contour lines.

- Large areas with no sounding data at all.

- Straight lines of soundings, rather than randomly spread across the chart.

- The words '**Unsurveyed**' or '**Inadequately surveyed**' or even 'Here be dragons' written on the chart.

- The letters '**ED**' (Existence doubtful), '**SD**' (Sounding Doubtful) or '**Rep**' (Reported, but not confirmed) against a chart feature.

If you see any of these features, you need to think about how much faith you are prepared to put in the charted information.

Circle 4

Circle 4 tells you that the tidal streams have been observed and predicted in the area covered by the chart. You will find the little magenta 'tidal diamonds' scattered around the chart in strategic positions. The predictions are referenced to an adjacent Standard Port and you are expected to interpolate between the neaps and springs figures to determine the tidal rate. This is covered more thoroughly in Chapter 13.

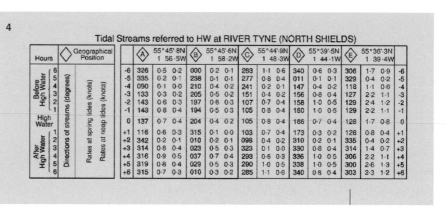

4

Tidal Streams referred to HW at RIVER TYNE (NORTH SHIELDS)

Hours	Geographical Position ◇	Directions of streams (degrees)	Rates at spring tides (knots)	Rates at neap tides (knots)	(A) 55°45'·8N 1 56·5W			(B) 55°45'·6N 1 58·2W			(C) 55°44'·9N 1 48·3W			(D) 55°39'·5N 1 44·1W			(E) 55°36'·3N 1 39·4W			
Before High Water -6					326	0·5	0·2	000	0·2	0·1	283	1·1	0·6	340	0·6	0·3	306	1·7	0·9	-6
-5					335	0·2	0·1	238	0·1	0·1	277	0·8	0·4	011	0·1	0·1	329	0·4	0·2	-5
-4					090	0·1	0·0	210	0·4	0·2	241	0·2	0·1	147	0·4	0·2	118	1·1	0·6	-4
-3					133	0·3	0·2	205	0·5	0·2	151	0·4	0·2	156	0·8	0·4	127	2·2	1·1	-3
-2					143	0·6	0·3	197	0·6	0·3	107	0·7	0·4	158	1·0	0·5	129	2·4	1·2	-2
-1					143	0·8	0·4	194	0·5	0·3	105	0·8	0·4	160	1·0	0·5	129	2·2	1·1	-1
High Water 0					137	0·7	0·4	204	0·4	0·2	105	0·8	0·4	166	0·7	0·4	128	1·7	0·8	0
After High Water +1					116	0·6	0·3	315	0·1	0·0	103	0·7	0·4	173	0·3	0·2	126	0·8	0·4	+1
+2					342	0·2	0·1	010	0·2	0·1	098	0·4	0·2	310	0·2	0·1	335	0·4	0·2	+2
+3					314	0·8	0·4	023	0·5	0·3	323	0·1	0·0	330	0·8	0·4	314	1·4	0·7	+3
+4					316	0·9	0·5	037	0·7	0·4	293	0·6	0·3	336	1·0	0·5	306	2·2	1·1	+4
+5					319	0·8	0·4	029	0·5	0·3	290	1·0	0·5	338	1·0	0·5	300	2·6	1·3	+5
+6					315	0·7	0·3	010	0·3	0·2	285	1·1	0·6	340	0·8	0·4	303	2·3	1·2	+6

Circle 5

Circle 5 is an example of the sort of more general warning that you can expect to find on charts. In this case, it draws attention to the various supporting publications that are available to a mariner to supplement the information on the chart itself. And it ends with the absolutely essential rejoinder to 'KEEP CHARTS AND PUBLICATIONS UP-TO-DATE AND USE THE LARGEST SCALE CHART APPROPRIATE.' Good advice.

5

IMPORTANT – SEE RELATED ADMIRALTY PUBLICATIONS
Notices to Mariners (Annual, Permanent, Preliminary and Temporary); Chart 5011 (symbols and abbreviations); The Mariner's Handbook (especially Chapters 1 & 2 on the use, accuracy and limitations of charts); Sailing Directions (Pilots); List of Lights & Fog Signals; List of Radio Signals; Tide Tables (or their digital equivalents).
KEEP CHARTS AND PUBLICATIONS UP-TO-DATE AND USE THE LARGEST SCALE CHART APPROPRIATE

7

Circle 6

There are a number of compass roses on any chart, aligned to true north with a smaller pointer to magnetic north. Written alongside the magnetic arrow, you will find the key for calculating the difference between true and magnetic north, which changes from year to year. We looked at this in more detail in Chapter 6.

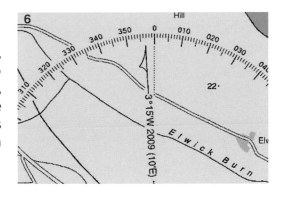

Other Things to Look Out for on a Chart

You haven't quite finished the initial inspection of the chart just yet. There are still a few more things to look out for:

Chart Correction State

The chart corrections that have been inserted are always listed at the bottom left-hand corner of the chart. They sit alongside the QR code that links you directly to a page on the UKHO website which tells you which corrections should have been applied to that particular chart. If possible, you should take the time to check this before you use the chart – so that you can trust the information that it is telling you. If you choose not to correct your chart, you will at the very least know by how much it is out of date.

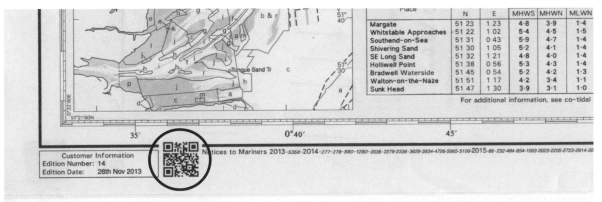

Chart Scale

Check that you are using the right scale of chart for your purposes. The UKHO, and most other chart publishers, produce mainstream charts in three generic scales. Ocean passage charts, on a small scale, are useful for long, open sea passages but contain nothing like enough information to safely approach the coast. Coastal charts, on a larger scale, contain more detailed information – enough to navigate safely in coastal waters, but without the precise and comprehensive detail that you need for inshore pilotage. And finally, ports, harbours and anchorages are covered by pilotage charts, which are the largest-scale charts, designed to contain all necessary information to get you safely in and out of harbour.

Self-evidently, you should always use the largest-scale chart appropriate to the navigation that you are doing at the time.

8 Depth and Elevations

A significant part of the chart-maker's art is the ability to give you an idea of the bottom topography at a glance.

On a metric chart, the shallowest waters (and therefore the most dangerous) are coloured in green and blue to make an immediate impression on the seaman's eye.

In addition, a significant amount of care goes into the selection and editing of the soundings in order to give you an uncluttered but unambiguous picture of the seabed.

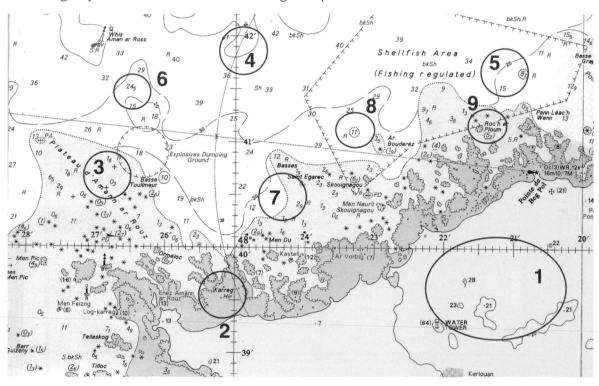

Circle 1

Land is always shown in yellow on a metric chart. From the chart-maker's perspective, 'land' is a place beyond the range of everyday tides, which is above the level of 'charted high water', or 'mean high water springs' (MHWS). In other words, it is anywhere above the line of seaweed that you find on the beach.

I was given a very useful tip when I was learning to be a professional navigator in the Navy. The line on your chart which represents

The black line usually corresponds to MHWS

8

MHWS, or the demarcation between dry land and the tidal area, quite often corresponds to the top of a thin band of black algae which discolours rocks at the high-tide mark.

If you are conducting precise navigation, this black line can be a useful vertical reference for measuring height, or as a horizontal reference for getting a precise fixing point on the edge of land.

The charted elevations of features above the high-water mark are generally measured from the level of MHWS, but they may use other datums and it is always worth confirming this in the Title Block of the chart. Spot heights are precisely plotted to mark hilltops that can be used for navigation. And Highest Astronomical Tide is generally used as the datum for measuring vertical clearance under an obstacle.

> ¹ **Heights** are in metres. Underlined figures are drying heights above Chart Datum; all other heights are above Mean High Water Springs.
> **Positions** are referred to the WGS 84 compatible datum, European Terrestrial Reference System 1989 Datum (see

Circle 2

The green area is the part of the seabed sandwiched between the level of Chart Datum and MHWS (i.e. the area of the chart which routinely covers and uncovers over the course of a tide). In this area, any soundings will be underlined – indicating that they are *above* Chart Datum, where they become 'drying heights', for example 3_5 indicates an area 3.5 metres above Chart Datum.

So, if you calculate a height of tide to be 6.2 metres, the *actual depth of water* at this spot will be 6.2 – 3.5 = 2.7 metres.

In practice some parts of these green areas will hardly ever be exposed, and other parts will seldom get wet, depending on their drying height and the tidal range.

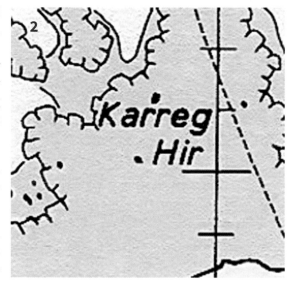

In my yacht, I always treat areas that are coloured green with extreme caution: if you don't need to sail over them – leave them well alone. Usefully, charts often show shallow areas that have a rocky seabed using a sort of cauliflower-shaped contour, while the smoother contours with gentle curves indicate a more even bottom, often sand or mud, which could provide a more relaxing spot to anchor.

Circle 3

Progressing into deeper waters, blue shading (sometimes light and dark blue) is used to delineate shallow waters that are below Chart Datum, out to a limit of 10 or 20 metres, depending on the

8

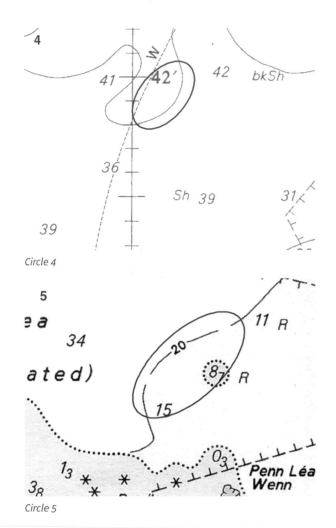

Circle 4

Circle 5

chart. On this chart, for instance, the dark-blue shading extends from 0 to 10 metres. The rest of the chart is unshaded.

Circles 4 and 5

This is quite important: note the difference between the complete contour line (Circle 4) and the broken line (Circle 5). When the cartographer has enough data to do so, he or she uses an unbroken line (Circle 4), and you can use this as a sign that he is more or less confident in the shape and the depth of the contour.

When he is more uncertain, however, normally owing to the quality or quantity of the sounding data, he uses a broken line (Circle 5). When you see these broken contours, you should sit up and think a bit: they are drawn for a good reason – to let you know that there is not enough accurate sounding data to draw a reliable contour line – and you should only use this part of the chart with very great care.

Circle 6

This is a simple sounding, indicating that the bottom lies 24.5 metres below Chart Datum. The actual position of the sounding is the centre of the area covered by the figures. Where the information is sufficiently robust to allow it, metric charts generally show depths of less than 21 metres to the nearest tenth of a metre: 16_8 indicating, for example, 16.8 metres below Chart Datum. That is to say that if you calculate the height of tide to be 3.7 metres, you will have a total depth of water of 16.8 + 3.7 = 20.5 metres. Charted depths between 21 and 31 metres, as here, are shown to the nearest half metre, and beyond that they are rounded *down* to a whole number (once again veering on the side of safety).

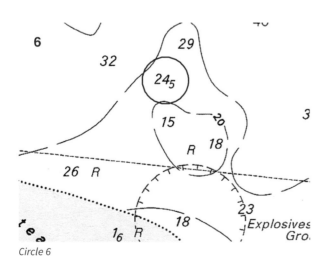

Circle 6

8

Circle 7

Diamonds may be a girl's best friend, but dotted lines do it for me. This dotted line is known as a 'danger line'. Your cartographer is only ever going to put a dotted line round something if he thinks it constitutes a 'danger to navigation', be that rocks, wrecks or, as in this case, a great swathe of the French coastline. Go there by all means, but be very well aware of the risks of doing so.

The actual words of *Chart 5011* (page 156 of this book) on this subject are:

> A danger line . . . delimits an area containing numerous dangers through which it is unsafe to navigate.

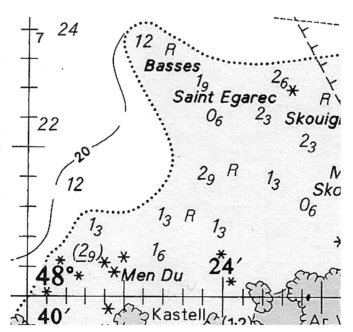

This is not quite 'Here be dragons', but it's not far off.

Circle 8

There are three pieces of important information contained in this one small area. First of all, the depth of the object is 11 metres below Chart Datum. Second, the symbol **R** indicates that it is a rock and, thirdly the dotted line round it shows that the chart-maker believes it to constitute a danger to navigation. Since the closest representative sounding is 25 metres, this rock stands about 14 metres clear of the seabed – a lot taller than the roof of my house.

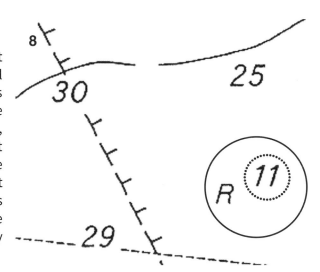

Circle 9

Here you have an isolated obstruction which is also standing well above the local seabed, to the extent that it actually dries periodically. If there is no space to print the sounding or drying height on isolated features such as this, the depth below, or height above, Chart Datum is shown close by, in brackets. This obstruction:

■ is called *Roc'h ar Ploum*

■ is a rock (because it is delineated by a cauliflower-shaped border)

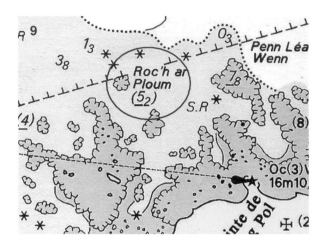

- stands above Chart Datum (because it is coloured green), and its height above Chart Datum is 5.2 metres

- and it stands on the edge of an area where fishing is regulated (the line consisting of inverted 'T' shapes).

Incidentally, you can safely assume that any obstruction in the middle of the sea which has been given a name by local fishermen must be significant – why else would they want to remember it? As a rule of thumb, all rocks are bad; but rocks with a name are often particularly dangerous.

Seabed Types

If you ever want to show off, learn the abbreviations for bottom types by heart, because I guarantee that you will be the only soul on board who has ever taken the trouble to do so. They might, of course, (and possibly with good reason) think that you should get out a bit more, but it would certainly be impressive. I will not go into all the abbreviations here. Sufficient to say that, if you

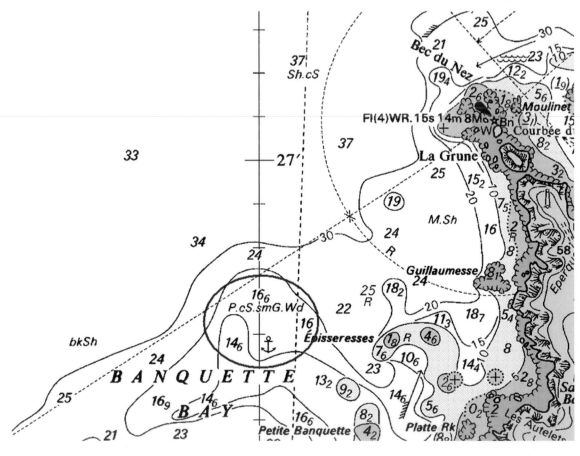

go to Section J of *Chart 5011* (page 153 of this book), you will find three pages of abbreviations for bottom type, adjective and colour.

Never again will you want for the official abbreviation for 'sticky green ooze' – or **sy.gn.Oz** as we navigators like to call it. I won't tell you what the abbreviation in the picture above stands for – just off the north-west coast of Sark in the Channel Islands; work it out, or look it up at the end of this book. You will notice, though, that it is close to a recommended anchoring spot, so the hydrographer has included it specifically to help vessels that are planning to anchor here. Bottom types are pretty rigorously documented and you would only have yourself to blame if you overlook these small, understated symbols and anchor on rock or some other treacherous part of the seabed.

Metric and Imperial Sounding

All soundings relate to Chart Datum. If you see **1**₃ on a metric chart, the sounding is 1.3 metres below Chart Datum. The same symbol on a fathoms chart means that the sounding is 1 fathom, 3 ft below Chart Datum – or 9 ft. Of course, in both cases, you need to add on the calculated height of tide to determine the actual depth of water at that spot.

Soundings can of course be qualified by the cartographer:

- **ED** stands for 'existence doubtful'
- **SD** for 'sounding doubtful'
- **Rep** for 'reported but not confirmed'.

A sounding with a straight horizontal line over it shows a 'safe clearance depth'. The precise depth may be unknown, but the obstruction is thought to be safe at the depth shown.

Equally, an underlined sounding in a kind of square saucer indicates a depth or obstruction that has been cleared by a wire sweep or diver. This symbol is used with other symbols like wrecks, rocks and wells.

Chart Datum

Just as all roads were once said to lead to Rome, all depths are rooted on that slightly elusive entity: Chart Datum. The trouble is that Chart Datum varies from spot to spot, depending on the behaviour of the tides: there is no line in the rock somewhere which acts as the ultimate arbiter of Chart Datum (unlike the UK Ordnance Datum, used for height above 'sea level' on all Ordnance Survey charts, which is carved into the harbour wall at Newlyn in Cornwall).

But Chart Datum is not too opaque either. As we have already established, it is generally set at or around the level of the lowest astronomical tide[1], and LAT for each of the Standard Ports is listed in the *Admiralty Tide Tables.*

---

[1] In some areas mean low water is used, and in non-tidal areas such as the Baltic, Chart Datum is generally taken as the mean sea level.

8

For most of us, the height of tide becomes pretty academic as you move offshore, but in some industries, like offshore energy extraction, it is of crucial importance. For that reason, the UKHO produces 'co-tidal charts' that illustrate variations in tidal height and timing across various stretches of water.[2] These co-tidal charts show the variations in tidal height and timing in offshore areas, which is profoundly useful to people building and operating oil rigs and other offshore structures. As such, they are pretty much designed for the specialist, rather than the general user, and most of us will only rarely come across them.

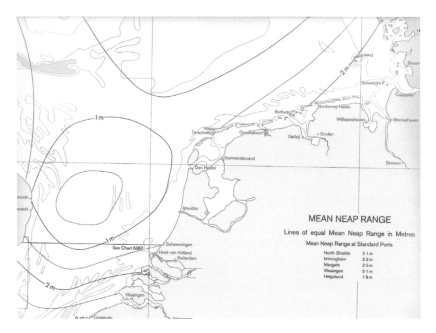

The chart above shows lines of equal tidal range while the second shows lines where the timing of low tide is the same. You will note that there is an extensive area about 25 miles east of Great

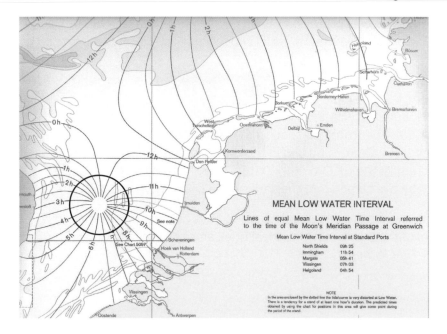

[2]UK waters, the Malacca Straits and the Persian Gulf.

8

Yarmouth where there is no rise and fall of tide at all. This is known in the trade as an 'amphidromic point' – a term of absolutely no use to the general navigator whatsoever, but an interesting physical phenomenon nevertheless. In the vicinity of one of these amphidromic points, the times of high and low water change rapidly around the circumference and you would have to calculate carefully if you are trying to achieve an accurate tidal prediction.

Charted Elevation and Height

The rules for charted height and elevation are really very simple:

■ On land, heights of hills, etc. are referred to either the MHWS or the local land datum (the Ordnance Datum in the United Kingdom).

■ The elevation of navigation lights are normally measured from MHWS.

■ Anything that you are going to pass underneath (like a bridge or power cables) needs to have an additional margin of safety built in, and is therefore measured from the level of the 'highest astronomical tide' (HAT).

These datums do vary from chart to chart but, as always, the chart's Title Block will give you clear details of which one is in use on that specific chart. See *Chart 5011* (page 130 of this book).

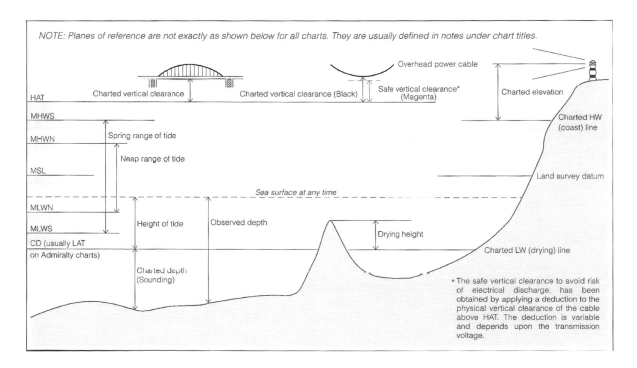

If you need to convert from one datum to another, you will find the necessary information either in the margins of the chart or in the local volume of the *Admiralty Tide Tables*.

8

How Far Away Can You See a Light or Other Feature?

There are three critical measurements for working out the range at which you can see a light.

The Nominal Range of a Light is the range at which, under normal night-time atmospheric conditions, with a meteorological visibility of 10 nautical miles, you would expect to pick up the light.

The Luminous Range of a Light is the range at which, under the prevailing atmospheric conditions, you would expect to pick up the light.

The Geographical Range of a Light, or another Feature is the range at which, in perfect visibility, it would disappear below the horizon. (Making allowance for normal conditions of atmospheric refraction.)

The distance at which you will be able to see a light will either be limited by atmospheric visibility or geography: it is therefore either the Luminous or the Geographical Range, whichever is less.

The distance at which you will be able to see a feature which is not illuminated is the Geographical Range or the meteorological visibility, whichever is the lesser.

The 'nominal range' of a light can be found either in the *Admiralty List of Lights*, or on the chart. In this case, Longstone Lighthouse's characteristics are shown on the chart overleaf as:

   Fl20s 23m 24M

that is to say: Flashing white every 20 seconds. The focal plane of the lens has an elevation above MHWS of 23 metres, and the light's nominal range is 24 nautical miles.[3]

The nominal range isn't of much practical use to you: what you actually need to do is to convert this nominal range into a 'luminous range', and you do this by looking in the *Admiralty List of Lights* for the 'luminous range diagram'[4], which will allow you to make this conversion.

So, the Longstone Lighthouse has a nominal range of 24 nautical miles (M). That is the range that you would see the light when the visibility was 10 M. How far away would you see the light if the visibility was 5 M? Come down from the top margin at 24 M until you hit the 5 M visibility curve and head off to the left-hand margin, where you will see the answer: 15 M. This is the luminous range.

However, you don't yet know whether you have sufficient height of eye to actually see the light at this luminous range, because the curvature of the earth may get in your way.

---

[3]My mnemonic is this: miles are bigger than metres, so Range (measured in miles) is given a capital 'M' and Elevation (measured in metres), is given a lower-case 'm'.

[4]The luminous range diagram is on page vii of the *Admiralty List of Lights.*

8

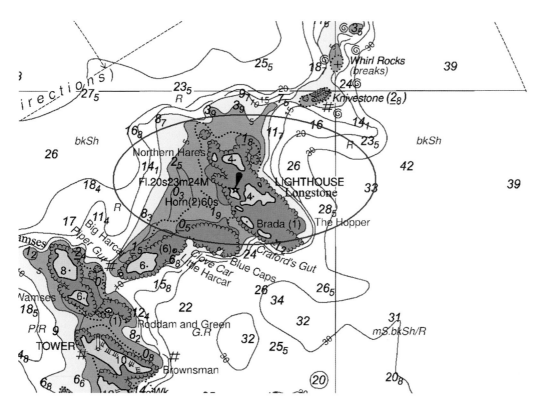

The Longstone lighthouse, showing light height and characteristics

LUMINOUS RANGE DIAGRAM

HD574

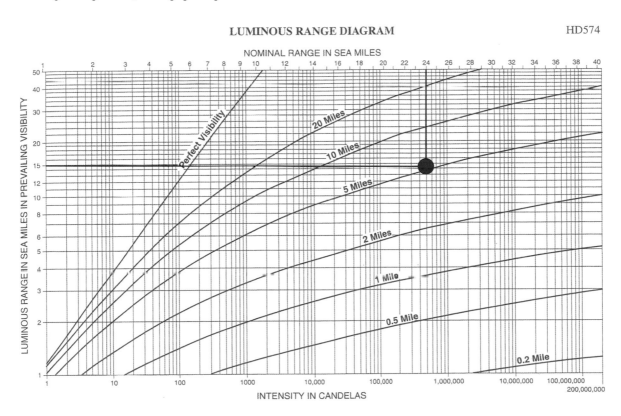

8

To work out the 'geographical range' of a light, or of any other object, you can look in the Geographical Range Table on page vi of the *Admiralty List of Lights*.

If the Longstone Lighthouse has an elevation of 23 metres, and your height of eye is 5 metres, you can expect to see the light, in perfect visibility, at 14.3 M: the geographical range.

Elevation in ft	m	Height of Eye of Observer in feet/metres																						
		3	7	10	13	16	20	23	26	30	33	39	46	52	59	66	72	79	85	92	98	115	131	148
		1	2	3	4	5	6	7	8	9	10	12	14	16	18	20	22	24	26	28	30	35	40	45
		Range in Sea Miles																						
0	0	2.0	2.9	3.5	4.1	4.5	5.0	5.4	5.7	6.1	6.4	7.0	7.6	8.1	8.6	9.1	9.5	10.0	10.4	10.7	11.1	12.0	12.8	13.6
3	1	4.1	4.9	5.5	6.1	6.6	7.0	7.4	7.8	8.1	8.5	9.1	9.6	10.2	10.6	11.1	11.6	12.0	12.4	12.8	13.2	14.0	14.9	15.7
7	2	4.9	5.7	6.4	6.9	7.4	7.8	8.2	8.6	9.0	9.3	9.9	10.5	11.0	11.5	12.0	12.4	12.8	13.2	13.6	14.0	14.9	15.7	16.5
10	3	5.5	6.4	7.0	7.6	8.1	8.5	8.9	9.3	9.6	9.9	10.6	11.1	11.6	12.1	12.6	13.0	13.5	13.9	14.3	14.6	15.5	16.4	17.1
13	4	6.1	6.9	7.6	8.1	8.6	9.0	9.4	9.8	10.2	10.5	11.1	11.7	12.2	12.7	13.1	13.6	14.0	14.4	14.8	15.2	16.1	16.9	17.7
16	5	6.6	7.4	8.1	8.6	9.1	9.5	9.9	10.3	10.6	11.0	11.6	12.1	12.7	13.2	13.6	14.1	14.5	14.9	15.3	15.7	16.6	17.4	18.2
20	6	7.0	7.8	8.5	9.0	9.5	9.9	10.3	10.7	11.1	11.4	12.0	12.6	13.1	13.6	14.1	14.5	14.9	15.3	15.7	16.1	17.0	17.8	18.6
23	7	7.4	8.2	8.9	9.4	9.9	10.3	10.7	11.1	11.5	11.8	12.4	13.0	13.5	14.0	14.5	14.9	15.3	15.7	16.1	16.5	17.4	18.2	19.0
26	8	7.8	8.6	9.3	9.8	10.3	10.7	11.1	11.5	11.8	12.2	12.8	13.3	13.9	14.4	14.8	15.3	15.7	16.1	16.5	16.9	17.8	18.6	19.4
30	9	8.1	9.0	9.6	10.2	10.6	11.1	11.5	11.8	12.2	12.5	13.1	13.7	14.2	14.7	15.2	15.6	16.0	16.4	16.8	17.2	18.1	18.9	19.7
33	10	8.5	9.3	9.9	10.5	11.0	11.4	11.8	12.2	12.5	12.8	13.5	14.0	14.5	15.0	15.5	15.9	16.4	16.8	17.2	17.5	18.4	19.3	20.0
36	11	8.8	9.6	10.3	10.8	11.3	11.7	12.1	12.5	12.8	13.2	13.8	14.3	14.9	15.4	15.8	16.3	16.7	17.1	17.5	17.9	18.8	19.6	20.4
39	12	9.1	9.9	10.6	11.1	11.6	12.0	12.4	12.8	13.1	13.5	14.1	14.6	15.2	15.7	16.1	16.6	17.0	17.4	17.8	18.2	19.1	19.9	20.7
43	13	9.4	10.2	10.8	11.4	11.9	12.3	12.7	13.1	13.4	13.7	14.4	14.9	15.4	15.9	16.4	16.8	17.3	17.7	18.1	18.4	19.3	20.2	20.9
46	14	9.6	10.5	11.1	11.7	12.1	12.6	13.0	13.3	13.7	14.0	14.6	15.2	15.7	16.2	16.7	17.1	17.6	18.0	18.3	18.7	19.6	20.4	21.2
49	15	9.9	10.7	11.4	11.9	12.4	12.8	13.2	13.6	14.0	14.3	14.9	15.5	16.0	16.5	17.0	17.4	17.8	18.2	18.6	19.0	19.9	20.7	21.5
52	16	10.2	11.0	11.6	12.2	12.7	13.1	13.5	13.9	14.2	14.5	15.2	15.7	16.2	16.7	17.2	17.7	18.1	18.5	18.9	19.2	20.1	21.0	21.7
56	17	10.4	11.2	11.9	12.4	12.9	13.3	13.7	14.1	14.5	14.8	15.4	16.0	16.5	17.0	17.4	17.9	18.3	18.7	19.1	19.5	20.4	21.2	22.0
59	18	10.6	11.5	12.1	12.7	13.2	13.6	14.0	14.4	14.7	15.0	15.7	16.2	16.7	17.2	17.7	18.1	18.6	19.0	19.4	19.7	20.6	21.5	22.2
62	19	10.9	11.7	12.4	12.9	13.4	13.8	14.2	14.6	14.9	15.3	15.9	16.5	17.0	17.5	17.9	18.4	18.8	19.2	19.6	20.0	20.9	21.7	22.5
66	20	11.1	12.0	12.6	13.1	13.6	14.1	14.5	14.8	15.2	15.5	16.1	16.7	17.2	17.7	18.2	18.6	19.0	19.4	19.8	20.2	21.1	21.9	22.7
72	22	11.6	12.4	13.0	13.6	14.1	14.5	14.9	15.3	15.6	15.9	16.6	17.1	17.7	18.1	18.6	19.1	19.5	19.9	20.3	20.7	21.5	22.4	23.2
79	24	12.0	12.8	13.5	14.0	14.5	14.9	15.3	15.7	16.0	16.4	17.0	17.6	18.1	18.6	19.0	19.5	19.9	20.3	20.7	21.1	22.0	22.8	23.6
85	26	12.4	13.2	13.9	14.4	14.9	15.3	15.7	16.1	16.4	16.8	17.4	18.0	18.5	19.0	19.4	19.9	20.3	20.7	21.1	21.5	22.4	23.2	24.0
92	28	12.8	13.6	14.3	14.8	15.3	15.7	16.1	16.5	16.8	17.2	17.8	18.3	18.9	19.4	19.8	20.3	20.7	21.1	21.5	21.9	22.8	23.6	24.4

So:

The Geographical Range is: 14.3 M

The Luminous Range in 5 M visibility is: 15 M

You can expect to see the light at the lesser of the two: 14.3 M

An Alternative

That's fine if you have a copy of the *Admiralty List of Lights* handy, but for leisure and small craft sailors there is a slightly less accurate way of calculating the geographical range of a light or an object. This entails getting your hands on a good almanac which has a table entitled **'Distance to the sea horizon'**.

You enter the table with your height of eye and find the distance to the sea horizon to get Distance 'A', and do the same for the other object to get Distance 'B'. Add the two together and, barring any severe atmospheric refraction, you have calculated the geographical range.[5]

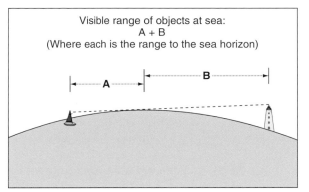

Visible range of objects at sea:
A + B
(Where each is the range to the sea horizon)

HEIGHT (m)	Distance to the horizon (M)	HEIGHT (m)	Distance to the horizon (M)
1	2.1	21	9.5
2	2.9	22	9.8
3	3.6	23	10.0
4	4.1	24	10.2
5	4.7	25	10.4
6	5.1	26	10.6
7	5.5	27	10.8
8	5.9	28	11.0
9	6.2	29	11.2
10	6.6	30	11.4
11	6.9	31	11.6
12	7.2	32	11.8
13	7.5	33	12.0
14	7.8	34	12.1
15	8.1	35	12.3
16	8.3	36	12.5
17	8.6	37	12.7
18	8.8	38	12.8
19	9.1	39	13.0
20	9.3	40	13.2

---
[5]You can also apply this technique to estimate the range of a vessel or shore feature that is hull-down over the horizon.

8

So once again, if the Longstone Lighthouse has an elevation of 23 metres, and your height of eye is 5 metres,

> A = 4.7 M
>
> B = 10 M
>
> The geographical range of the light is 14.7 M.

This isn't as accurate as using the geographical range tables, but it isn't far out.

In many cases, you can get an approximate range at first sighting by comparing this range (14.7 M) with the nominal range of the light (24 M), making a mental allowance for the actual visibility, and taking the lesser of the two.

Of course, with a strong lighthouse on a clear night, you may be able to see the loom of the light well before you actually see the light itself.

9 Landmarks, Lights and Coastal Features

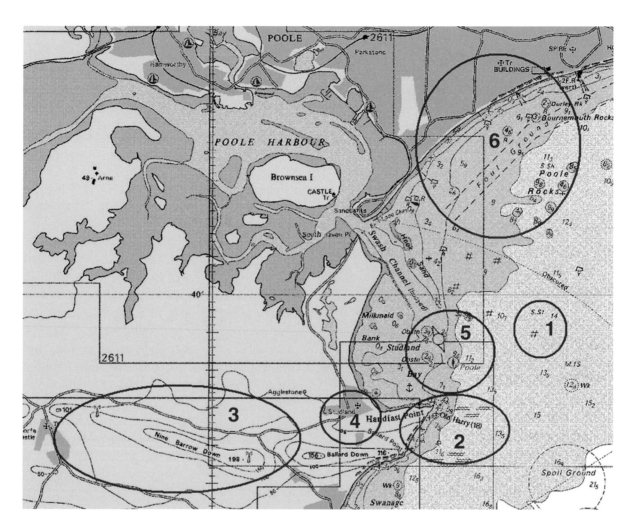

The only reason for a chart to include detail of land features behind the coast is to assist the mariner with navigation, or to help him or her understand the nature of the hinterland a bit more clearly. It features plenty of useful information for the mariner.

Circle 1

This is a chart of Studland Bay, a well-known and popular anchorage on the south coast of England. The bottom type **S St** indicates sand and stones – good for anchoring. It does have a few hazards, though: the hash sign # indicates a bottom obstruction that you need to be aware of; it is not a danger to navigation (if it were, it would have a circle of dots around it), but you would not want to drop your anchor on top of it.

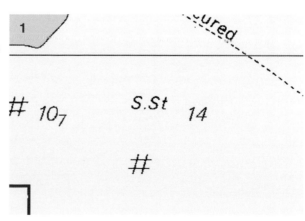

9

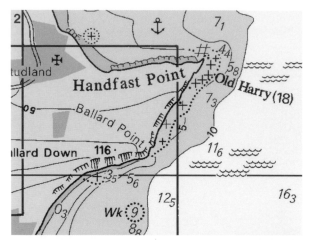

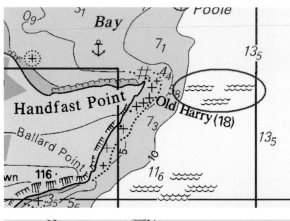

Old Harry Rock, taken from Studland Bay, looking south-east. Note the contrast between the calm water of the bay and the overfalls beyond the headland

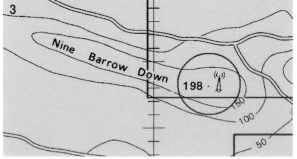

Circle 2

Another good indicator of a sandy or muddy bottom is the languid curve of well-separated bottom contours northwest of Circle 2, in contrast with the very much steeper shelving bottom by Ballard Point, to the south of Old Harry, where cliff erosion has created a rocky and potentially dangerous seabed. The 'eyelashes' drawn round Ballard Point show a cliffy part of the coast, which often implies a shelving, rocky seabed beyond.

Whenever you are navigating close to a sharp point of land in tidal waters, you should always expect there to be some uncomfortable water off the tip: the Point of Ayr on the Isle of Man, Portland Bill, St Catherine's Point, the Raz de Sein. These races can sometimes be dangerous to larger ships, let alone small fishing boats or yachts, particularly when spring tides are running and the wind is blowing against the tide. Handfast Point is no different, and its race has been marked on the chart with a series of ripples extending half a mile or so to seaward. This is a chart symbol to take seriously: if I was planning to pass through an area of charted overfalls, I would always read the local pilot or *Sailing Directions* beforehand in order to find out how severe the race was likely to be – and what conditions should specifically be avoided.

Circle 3

Moving inland, there are two radio masts on Nine Barrow Down, which could presumably be used for visual fixing. The topography of these hills has been shown in some detail; in particular, the peak and the slightly lower one on Ballard Down have been precisely plotted, and you might consider using them as aids to navigation.

Circle 4

In the village of Studland itself, the church provides another fixing mark. This may or may not be useful to the mariner, depending on its visibility, but in general, churches are hugely useful. Speaking as a long-time navigator, I would recommend that the Church Commissioners build all of their coastal churches with a good view of the sea, well clear of trees and other housing, and with tall, well-lit spires.

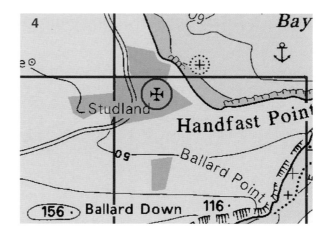

Major roads in coastal areas are generally set out only in fairly sketchy detail: don't expect the road network to be laid out as accurately or as fully in a chart as it is in an Ordnance Survey map. Roads can very occasionally be used for navigation when they run directly down a slope towards the sea, although I have tried this on a number of occasions without great success. They do, however, serve to illustrate how the various land features link up.

Circle 5

There are a variety of symbols here. The small magenta circle containing a solid vertical diamond shows the location of the Poole Harbour pilot station: if you are going to take a pilot for passage into the harbour, or drop him off after leaving, this is where you do it. The other circle, with two triangles attached, represents the radio reporting point for traffic proceeding in and out of the harbour (you need to refer to the *Sailing Directions* for more

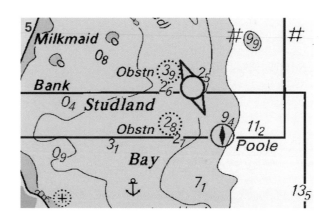

details of who you must call, under what conditions and on what frequency). There are also two bottom obstructions in this circle, 2.8 and 3.9 metres below Chart Datum respectively. These have each been given a dotted line 'necklace' as a warning that they should be considered a danger to surface navigation.

Circle 6

There are a multitude of bottom obstructions: a warning to mariners seeking to anchor in this part of the bay, which would otherwise be an ideal spot to shelter from a westerly or north-westerly blow. There is a long thin patch of foul ground and a big legend saying 'Poole Rocks', not to mention a number of underwater pipes sticking out from the coast.

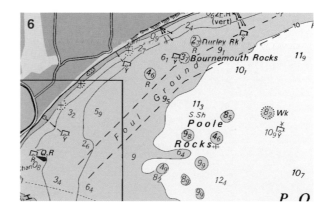

9

Beware Trees, Coastal Topography and Property Developers!

One element of successful pilotage is to take time and have a good look at the chart before you enter harbour and pick out a succession of landmarks which will help guide your passage. But beware: trees grow up in front of navigation marks or churches (particularly, it seems, in French coastal towns) and landmarks have an uncanny way of hiding themselves just when you need them most.

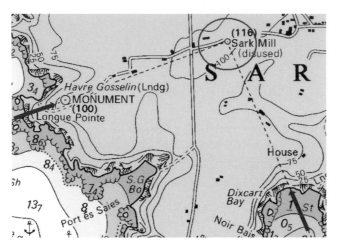

One of the classic cases of 'The Great Disappearing Landmark' is the Sark windmill, perched right on the top and centre of the Island of Sark, in the Channel Islands. The hydrographer, with his irrepressible sense of humour, suggests that you can use the disused mill as a back mark of a transit when approaching from the south-east or south-west: these transits are drawn as deliberate lines on the chart. And of course you can use them ... but only up to a point, because, as you will have spotted, the mill has a charted elevation of 116 metres, and the cliff edge has an elevation of about 75 metres. So it will at some stage disappear right behind the cliffs, just about as you are closing the land – the point where you need this landmark the most.

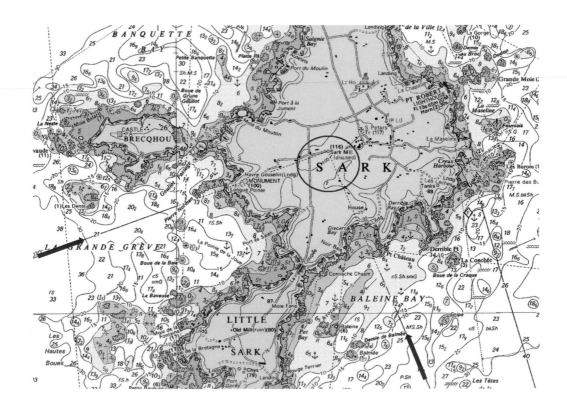

And then there are the eager property developers who build blocks of flats which hide landmarks and even sometimes block navigationally significant transits. One of the most striking of these in my experience is the transit between the Old War Memorial and St Jude's Church on the entry into Portsmouth. This is not too significant for big ships going up the channel, but for small boats coming in through the Swashway, it marks an important channel between two patches of shallow water. The relatively recent 'BUILDING' – a big, square block of flats – has been constructed precisely at the point where it obscures your view of St Jude's spire when you are on the transit. The obvious solution is to use the right-hand edge of this building as the back mark; although it is not as precise as its predecessor, it is quite adequate for getting you into the Swashway.

None of these little quirks of navigation is impossible, or even particularly difficult to deal with, but time spent quietly and intelligently studying the chart in advance will often pay dividends – and may remove a few unwarranted surprises – when you come to the actual business of navigation.

Navigation Lights and Buoys

Charts go to a lot of trouble to highlight navigation lights and spell out their characteristics. Navigation lights are universally marked by a little magenta 'flash' that sticks out at an awkward and rather uncomfortable angle from a small black star (in the case of fixed lights) and a black circle (for buoys).

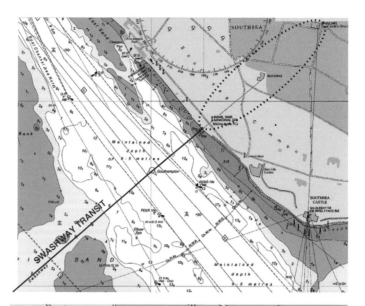

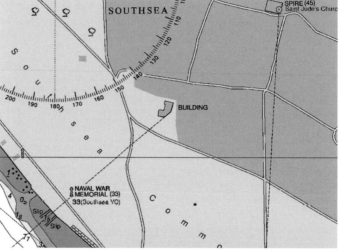

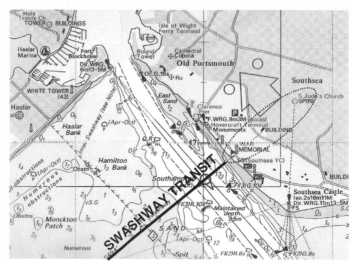

9

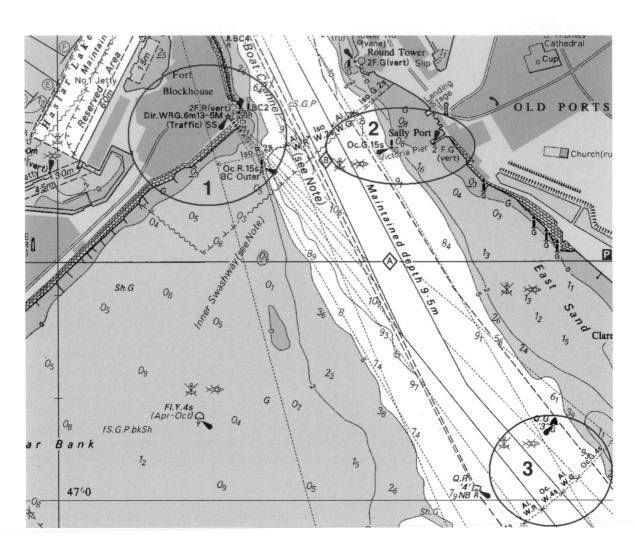

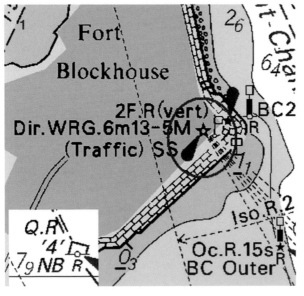

Circle 1

As you would expect, the entrance to Portsmouth Harbour is pretty well-lit. On the final leg into the harbour, the principal leading light is located on top of Fort Blockhouse. The specification reads: **Dir.WRG.6m 13-5M**. That is to say that the light is directional (i.e. it is only visible over certain sectors of arc), the sectors show white, red and green respectively, depending on the sector that you are in. The focal plane of the light is elevated 6 metres above mean high water springs (MHWS) and the nominal range of the light depending on sector varies from 13 to 5 nautical miles.

Circle 3

Further down the channel, you will see the sectors picked out – alternating white and red, occulting white every 4 seconds, alternating white and green, etc. This sectoring is a relatively simple way of indicating to the mariner where he is located in relation to the centreline of the channel, and as a result the light sectors are very precisely mapped out and executed. More details are contained in the *Admiralty List of Lights*.

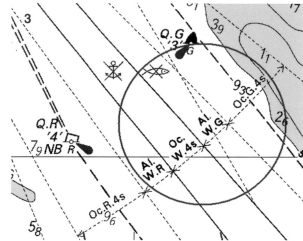

Close to the north-east of the principal leading light (Circle 1), and situated almost on the water's edge, are two fixed red lights, arranged vertically, one above the other, **2F.R(vert)**.

Circle 2

To the east of the entrance are two lit features: a pier and a beacon. The pier has two fixed green lights, arranged vertically (**2 F.G(vert)**), and the beacon has a green light occulting every 15 seconds (**Oc.G.15s**). Occulting is the absolute opposite of flashing – the light is on, except for periodic 'flashes' of darkness – the occult. You will see that the beacon also has a capital **G** to the right of it, which

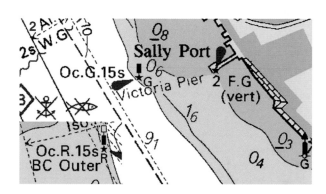

indicates that the beacon itself is painted green. This beacon has a counterpart called **BC Outer** on the port side of the channel (inset), which is painted red and occults a red light every 15 seconds.

Characteristics of Navigation Lights

If you look at Section P of *Chart 5011* (page 167 of this book), you will find the various different characteristics of navigation lights laid out in precise detail. In essence, there are eight ways to describe the flashing rhythm: Fixed, Occulting, Isophase, Flashing, Quick, Very Quick, Ultra Quick and Alternating. And, of course, they can be combined to form 'Very Quick Flashing', etc. Annoyingly, the symbols on UK charts are subtly different from those on international charts, but this is unlikely to cause confusion.

In each case, the period given is the time between the start point of one cycle and the same point on the next cycle.

9

Flashing rhythm	Description
Fixed	The light, which might be any colour, stays on continuously.
Occulting	The total period of light is longer than the total period of darkness. I remember this as a light which is predominantly 'on', flashing periods of darkness at regular intervals. May be: ■ Single-occulting – one period of dark in any cycle. ■ Group-occulting – more than one period of dark in a cycle, but gathered together as a group. ■ Composite group occulting – more than one period of dark in a cycle, but gathered together in more than one group.
Isophase	The duration of the light is equal to the duration of the darkness. A very regular rhythm with the light on for a period and then off for the same period. It stands out very easily from flashing or occulting lights.
Flashing	Total duration of the light is less than the total duration of darkness. In other words, precisely what it says on the tin. May be: ■ Single flash – one period of light in any cycle. ■ Group flashing – more than one period of light in a cycle, but gathered together as a group. ■ Composite group flashing – more than one period of light in a cycle, but gathered together in more than one group. ■ Long flash – where the length of the flash is 2 seconds or more.
Quick	A repetition rate of between 50 and 79 flashes per minute (although for simplicity it is normally either 50 or 60 flashes per minute). May be: ■ Continuous quick flash – no breaks in the sequence of flashes. ■ Group quick flashing – more than one flash, grouped together, in each cycle. ■ Interrupted quick flash – a continuous quick flash, but with a break between cycles.
Very Quick	As for a Quick Flash, with the same groupings, but with flashes repeating at a rate of 80 to 159 flashes per minute. Usually either 100 or 120 flashes per minute.
Ultra Quick	A light with flashes repeating at a rate of not less than 160 flashes per minute.
Alternating	The light alternates between two colours, each exposed for the same length of time, and with no periods of darkness between them. In Circle 3 of the Portsmouth lights diagram, for instance, you will see that the directional navigation lights alternate white and green, or white and red, depending on where you are relative to the centreline of the channel.
Morse Code	A light which shows groups of long and short flashes which are designed to represent characters of the Morse Code. For instance, a light which is marked 'Mo(K)W' would flash the morse code for 'K', or '— - —' using a white light.

Directional Lights

In many places, like the entrance to Portsmouth, or on a lighthouse standing guard over a shoal area, fixed lights show different colours or characteristics over different sectors. On the chart, these sectors are very often shown as dotted lines radiating outwards from the light. Be careful, though, because in *Notices to Mariners,* and in the *Admiralty List of Lights*, the limits of each sector are *written down from the mariner's perspective*: as 'the True bearing of the light, seen from seaward'. A common and potentially dangerous error is to plot these bearings from the light outwards. This would give a mirror image and prove quite puzzling on a dark and stormy night.

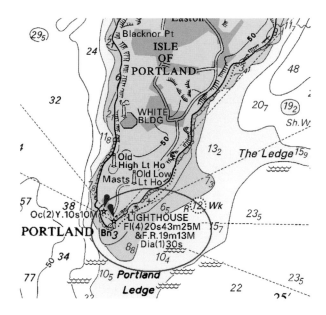

The main light characteristic of Portland Bill lighthouse is a white light Flashing (4) every 20 seconds. This is supplemented by a solid red light that shows over a 20° sector and covers the Shambles Shoal to the east-south-east of the light.

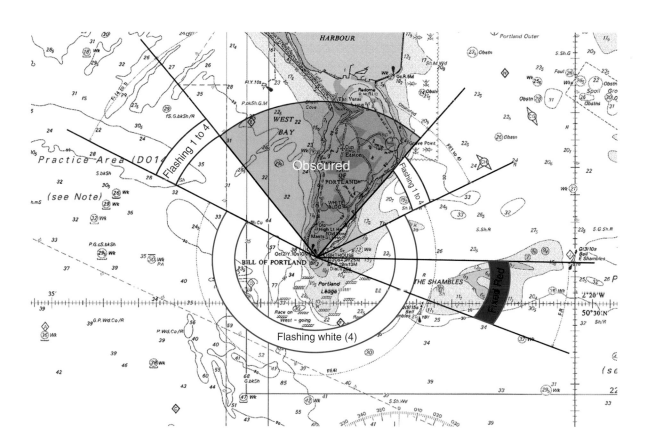

9

In the north-west and north-east sectors, the white light's characteristics change as you move northwards, gradually reducing from Flashing (4) to Flashing (1). *The Admiralty List of Lights* contains more detailed information than is shown on the chart, including the colour scheme of the lighthouse itself.

The **Elevation** of the light (in metres) is shown in the characteristics with a lower-case 'm' after it, alongside the characteristic of the light. This is normally the height of the focal plane of the light above MHWS. It sounds obvious, but it is worth bearing in mind that, on nights when there may be fog patches or low cloud, high-elevation lights, especially those situated on top of cliffs, may be completely obscured even though sea level visibility is adequate.

A0294	- Portland (TH)	50 30.85 N 2 27.38 W	Fl(4)W 20s	43 41	25	White round tower red band	(*fl* 0.1, ec 1.4) x 3, *fl* 0.1, ec 15.4. Gradually changes from 1 *fl* to 4 *fl* 221°-244°(23°) 4 *fl* 244°-117°(233°). Gradually changes from 4 *fl* to 1 *fl* 117°-141°(24°)
	..	F R	19	13	..	Vis 271°-291°(20°) over The Shambles
...	..	Dia 30s	..	.. ..		*bl* 3.5

And One Final Word of Advice

Be very careful when you are trying to identify navigation lights against a backdrop of shore lighting. All manner of street lights, fun fairs, pubs and coloured lighting along the seafront have been known to impersonate navigation lights. I was once greatly embarrassed by a fixed green light which I was happily using for navigation . . . until it suddenly turned red. This was not the solid and dependable navigation light that I had so confidently supposed, but a railway signal light . . . and my colleagues never let me forget it! Sometimes, too, in situations where there is a lot of shipping around, a fixed light may be briefly obscured by a ship or obstruction, making it appear as if it is flashing. You must never make assumptions based on what you would like to see: if it doesn't feel right, check again.

Sound Signals Associated with Navigation Marks

If you are ever worried that you may just be feeling too happy, go down to the coast and sit within earshot of a fog signal for 10 minutes. They are universally the most depressing sound ever conceived by mankind. There is a sort of desolate loneliness about them which must have been devastating for mariners in the age before precision navigation and radar – and quite frankly they are bad enough today. No wonder they gave up manning lighthouses.

Fog signals come in seven flavours: explosive, diaphone, siren, horn, bell, whistle and gong. These sounds are all pretty self-explanatory, with the exception of a diaphone, which is, to the laymen among us, essentially two air-powered whistles going off at the same time to make a rather puffy, discordant sound like the steam trains in Western movies used to make.

Sound signals can be produced from lighthouses, light ships, buoys or other navigation marks, and they are, where appropriate, given both a periodicity and an identity, for example:

- **Horn(1) 15s**: a sound signal which gives one blast of the horn every 15 seconds

- **Siren Mo(N) 60s**: A siren which makes the Morse Code for 'N' (━ •) every 60 seconds. ('Mo' is the abbreviation for 'Morse')

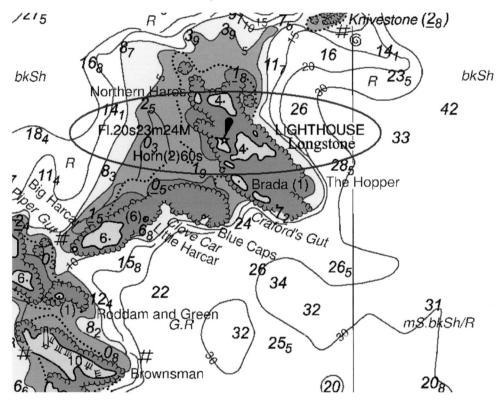

Returning briefly to Longstone Lighthouse on the Farne Islands, you will see the light (flashing every 20 seconds, etc.) described on the top line, and its associated sound signal (**Horn (2) 60s**) on the bottom line. This signal sounds two blasts on a horn every 60 seconds. You are only likely to hear this signal at times of restricted visibility.

You really cannot predict the range of a sound signal: it depends on aspect, wind, intervening obstructions and, any number of other factors.

9

Buoys and Beacons

A buoyage system is a country's way of guiding unfamiliar mariners through dangerous waters. Its meaning must, therefore, be completely unambiguous both on the water and on the chart. The IALA[1] System was developed in the late 1970s and, with a few well-documented exceptions, has now been adopted around the world. It consists of five complementary types of mark: lateral, cardinal, isolated danger, safe water and special marks, which may be used in any combination. These marks can be floating or shore-based, buoys or beacons, and they come in a variety of shapes and sizes, many of which are distinguished on the chart.

Lateral	Green and red, according to the side of the marked channel	Used to indicate the limits of a navigable channel.
Cardinal	Yellow and black in horizontal bands, depending on the direction of the obstruction	Indicates the direction, in reference to a compass, along which the danger lies. A northerly cardinal mark lies to the north of an obstruction, and ships in turn should pass to the north of the mark.
Isolated Danger	Red and black in horizontal bands	Shows isolated dangers of limited size with navigable water all round.
Safe Water	Red and white in vertical stripes	Shows that there is safe and navigable water all round. Used for landfall and mid-channel marks.
Special	Yellow	No navigational purpose: used to mark a feature like a sailing mark or the end of an outfall pipe, etc.

I do not propose to labour the particular features of cardinal and lateral marks here: this is essential knowledge of any mariner in any form of vessel and I will assume that you know it.

Cardinal marks are referred to the cardinal points of the compass; lateral marks are referred to the direction of flow of the flood stream. It is important, therefore, to ensure that, when using lateral marks, the buoyage authorities and the mariner are making the same assumption for the direction of the flood. Where appropriate, or where there may be doubt, charts are marked with a hollow magenta arrow pointing in the direction of the flood (known as the 'direction of buoyage'). However, there are a few critical points around the coast where the assumed direction of buoyage changes abruptly. One of them is shown on the chart opposite.

Here, in the Menai Straits in north-west Wales, the precise point of this change direction of buoyage changes is marked by a buoy called 'Change', just off the city of Caernarfon. See how the hollow arrows indicating the direction of buoyage meet nose-to-nose at this point.

If you look carefully, you can see the different-coloured buoys marking the north side of the channel above and below the 'Change' buoy.

-----

[1]IALA stands for the International Association of Lighthouse Authorities.

9

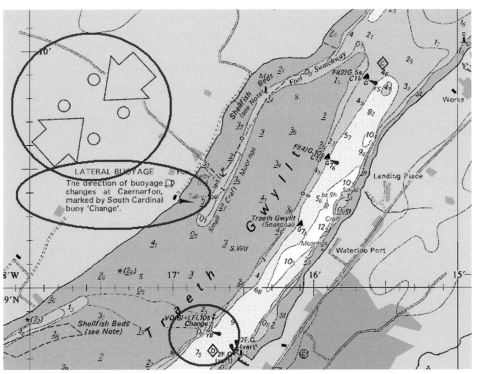

Chart of the Menai Straits, Wales

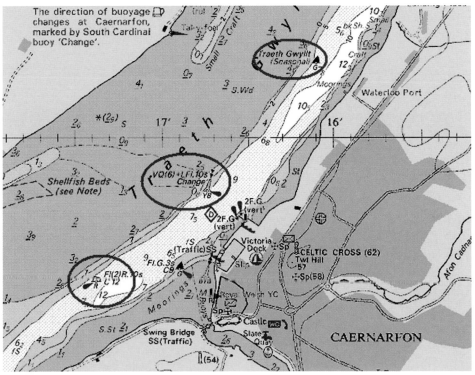

Showing how direction of buoyage can change

Navigators should also be aware that there are two systems of lateral marking in use around the world, divided into two regions: 'A' and 'B'. Region A includes Europe and the Mediterranean, and required red marks to be left to port and green to starboard when moving with the flood stream. Region B operates in the Americas and areas of the West Pacific, where the colours are reversed: red to starboard and green to port when proceeding in the direction of the flood stream.

An Example from the Western Solent

The western approach to the Solent, at the Needles Channel, shows a variety of very distinctive buoys, all with different characteristics and all giving discrete messages to the mariner.

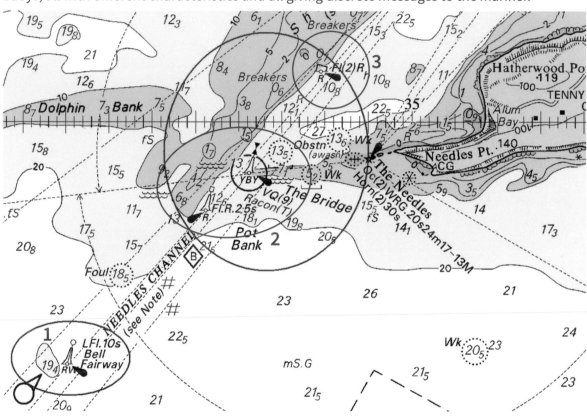

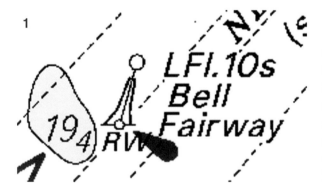

Circle 1

As you approach from the south-west, the first buoy that you encounter is the Fairway Buoy, a 'safe water mark' with navigable water all round it, acting as the principal landfall marker. It shows a long flash of white light every 10 seconds, has a bell and is painted white and red in vertical stripes (as indicated by the vertical line down the centre of the buoy symbol and RW underneath). It has a circular topmark.

9

Circle 2

You are now in the Needles Channel, moving in the direction of the flood tide, and the next mark that you encounter is the red lateral mark, which flashes red every 2.5 seconds. It is painted red and has a cylindrical topmark. You leave this mark to port.

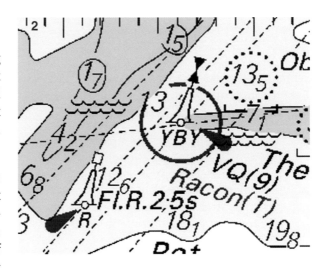

Shortly after that, you come across a westerly cardinal mark (which you must keep to the west of). It is painted yellow-black-yellow in horizontal bands, has a westerly cardinal topmark consisting of two cones apex together and a light that quick-flashes in groups of nine. You will notice that this buoy also has a magenta circle round it on the chart, and **Racon (T)** written in magenta underneath. That indicates a radar transponder which, when interrogated, displays the Morse symbol 'T' (━) on your radar display to assist identification.

Circle 3

Moving further north-east into the channel proper, you start to encounter the Western Solent's lateral buoyage system of red and green buoys. Since you are travelling in the direction of buoyage (i.e. the direction of the flood stream), the red buoys should pass down your port side and the green buoys to starboard.

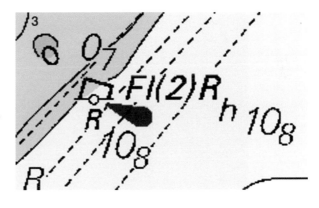

10 Dangers: Wrecks, Rocks and Obstructions

Wrecks and rocks can stand out a very long way from the seabed and they are pretty unforgiving. Even though modern charts will always try to show a least depth over underwater obstructions, you should always be aware that:

■ The survey ship may not always have been equipped to find the least depth over a danger, particularly if the survey was undertaken before the mid-1970s.

■ A wreck may have shifted since the most recent survey or, in areas where the seabed is unstable, the depth may well change by a considerable margin in a short space of time. The town of Pozzuoli on the north-western coast of the Bay of Naples is a good example of this. A volcanic effect known as 'bradyseism' has caused significant and rapid vertical movement of the seabed over the last few thousand years. The most recent events occurred between 1968 and 1972, when the whole town, port and seabed rose by 1.7 metres, followed about 10 years later by a second rise of 1.8 metres.

■ Or that the position may not always be accurately charted.

Modern surveys, sidescan and later systems[1], are pretty good at finding obstructions and accurately calculating the depth of water over them: if you are in an area where the charting derives from an earlier survey, you should if possible give these obstructions a wide berth.

Wrecks

Multi-beam sonar image of the wreck of a Norwegian freighter sunk in the Dover Strait

On any chart in general use, a wreck will be annotated with the cartographer's most accurate information of the least depth of water over it. Some, depending on the depth of water, will be classified as 'dangerous to surface navigation'. The criterion of 'dangerous' has increased over

---

[1]Sidescan sonar came into general use in British survey vessels in 1973.

10

the years (see Chapter 3), but any chart drawn after 1968 categorises any wreck with less than 28 metres of clearance as 'dangerous to surface navigation' – and it will be drawn on the chart with a dotted 'necklace' around it.

Any semi-competent mariner needs to be able recognise the different wreck symbols that are shown on his chart, and explained in Section K of *Chart 5011* (page 156 of this book). These aren't particularly difficult to learn – more significant is the message that they are giving you over the level of risk. In general, information on a wreck which has been surveyed by modern surveying techniques (since about 1973) is likely to be accurate, as will the information on a chart that has been swept by wire. A wreck that has only been surveyed by echo sounder in a survey pre-dating the advent of sidescan sonar may, however, be less accurate.

It goes without saying that the depth of the wreck, if less than 5 or 10 metres (depending on the convention of that particular chart) will be shaded blue as an additional caution to the mariner.

If you look at this chart of the north-east coast of Brittany, you will see a number of relatively ugly obstructions scattered across the approaches to Saint-Quay-Portrieux.

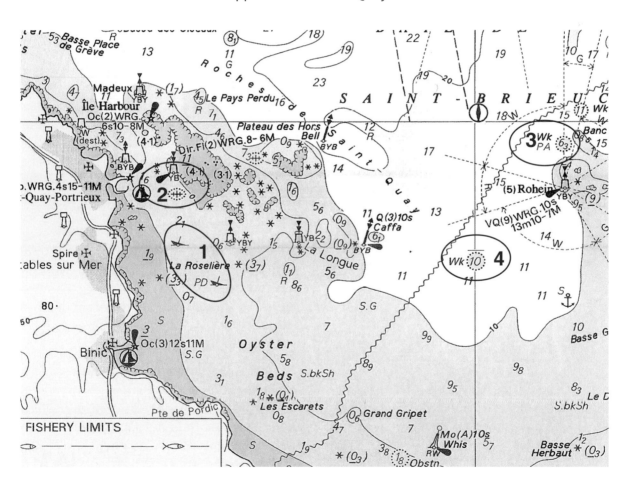

10

Circle 1: wreck that shows some of its masts or superstructure at Chart Datum

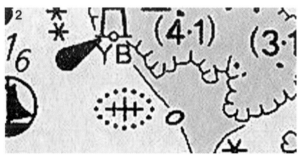

Circle 2: wreck of unknown depth that is considered to be a danger to surface navigation

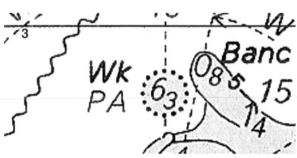

Circle 3: wreck with 6.3 meters of water over it, a danger to surface navigation, position approximate

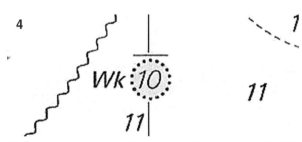

Circle 4: a wreck over which the exact depth is unknown, but which is considered to have this amount (10 metres) of water over it

Circle 1

To start with, there are two wrecks that show their hull or superstructure at Chart Datum (you may of course not see anything at all, depending on the state of the tide). The more southerly wreck is marked **PD** (position doubtful). Give it a wide berth, even so.

Circle 2

Next, there is a wreck of unknown depth which is considered a danger to navigation and is coloured blue to indicate that, in the cartographer's view, the wreck is less than 10 metres below Chart Datum. This triple-cross symbol represents a wreck of unknown depth, or – self-evidently not in this case – one with a depth of more than 200 metres.

Circle 3

This shows a wreck with 6.3 metres of water over it at Chart Datum, but whose position is only approximate. This is also considered to be a danger to navigation.

Circle 4

Shows a wreck, 10 metres below Chart Datum, with a single black horizontal line over the symbol. This line indicates a 'safe clearance depth': a wreck over which the exact depth is unknown, but one that is considered to have this safe depth of water over it.

And finally, there is one category of wreck which does not appear on that chartlet, which you still need to be able to identify: a wreck, over which the precise depth is not known, but which has been cleared to the given depth by a wire sweep. This one is also a danger to navigation.

10

Rocks

Unlike wrecks, there is no convention of minimum depth when deciding whether a rock is dangerous to surface navigation (or any other kind of navigation for that matter). Individual rocks, or groups of them, are enclosed in a dotted 'necklace' at the hydrographer's discretion. Speaking for myself, though, I would consider each and every rock a danger to navigation unless I had clear and positive evidence to the contrary.

Have a look at this chart. It is the entrance to one of my favourite yachting destinations, a port called L'Aber-Wrac'h on the north-western coast of Brittany, France. This port is interesting not because of the shore facilities but because of the sheer excitement of weaving your way through the rocks that lie behind the impressive lighthouse of the Île Vierge. This is a wrecker's coast if ever there was one.

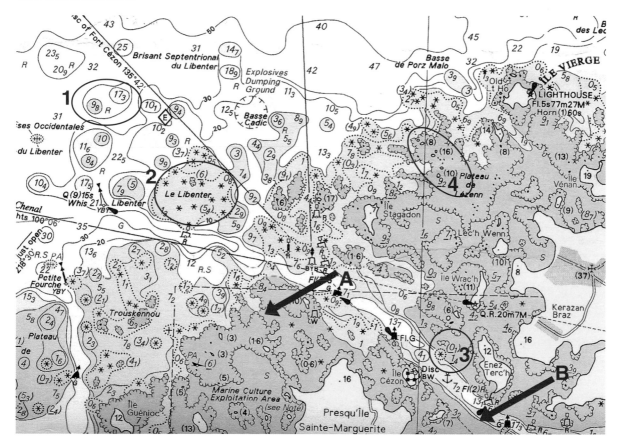

L'Aber-Wrac'h is to hull-munching rocks what Crufts is to pedigree dogs, yet much of the seabed between rocks, just like the Cornish coast, has a sandy bottom. You can see how the different bottom types are illustrated on the chart, with the 'cauliflower' outline (**Arrow A**) delineating areas of rocky bottom with sand in between, and the smoother, more predictable shape (**Arrow B**) where the bottom is more fluid: sand or mud.

As you would expect, this area contains much of the UKHO's lexicon of rock types, coloured green where they dry between Chart Datum and MHWS, blue (two shades) where the depth of the rock is between 0 and 10 metres, and white where the depth is greater than 10 metres.

10

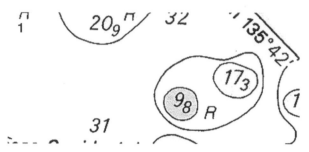

Circle 1: a pair of rocky pinnacles, standing about 20 metres proud of the seabed

Circle 1

In Circle 1, you will see a pretty classic couple of rock pinnacles coming up steeply from the bottom to a depth of 9.8 metres and 17.3 metres respectively. The shallower of these two stands about 20 metres proud of the surrounding seabed (about the height of a five-storey building). You know that these are rocky outcrops because of the symbol **R**, and because of the steepness of the contours that would have been eroded in the tidal streams long ago if the pinnacles were made of any other material. Looking around, you will see a number of other pinnacles, which is pretty typical of this sort of coast. In general, pinnacles seldom stand alone: if the geological conditions are right for one, there will often be others.

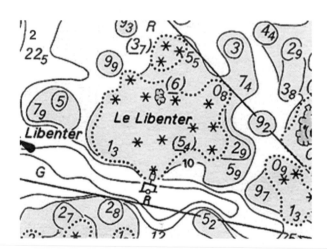

Circle 2

A short distance to the south-east of Circle 1, you will find the Libenter Bank (Circle 2), which is an absolute bear-pit of rocks. The whole area is surrounded by a dotted necklace to indicate that this is not somewhere to take a vessel without serious consideration. There are a multitude of little black star shapes. These indicate a single rock which covers and uncovers, in other words its peak lies somewhere between Chart Datum and MHWS. They sometimes, but not always, have a drying height associated with them. One of the rocks in this group, for instance, has a drying height of 5.4 metres. Where the drying height is set aside from the rock, rather than superimposed upon it, the height is shown in brackets. Rocks which are larger than a 'single rock' become a cluster, and they are given their own cauliflower shape. There is one rock cluster in Circle 2, coloured green, with a drying height of 6 metres above Chart Datum.

Circle 3

Rocks don't always come in multi-buy packs. In Circle 3, there is a single rock lurking on the side of the main fairway into the port. This rock, which is considered a danger to navigation, covers and uncovers and has a drying height of 0.1 metres.

Circle 3: single rock, close to the main fairway, which is a danger to navigation. Standing 0.1 metres above Chart Datum

10

Circle 4

The most prominent rocks are those which never cover – small islets, in other words, standing higher than MHWS. A cluster of about seven is shown in Circle 4, small dots of yellow against the green background, standing 8 to 16 metres above Chart Datum and well above the rock base, which has a drying height of 9.2 metres.

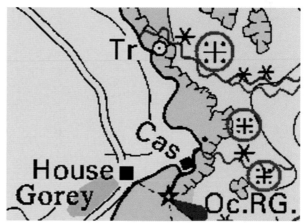

Circle 4: a group of about seven small islets that stand above MHWS

You should, of course, always check the Title Block of your chart to make sure that you understand which datum is being used for the height of objects above the water.

Depths are in metres and are reduced to Chart Datum, which is approximately the level of Lowest Astronomical Tide. **Heights** are in metres. Underlined figures are drying heights above Chart Datum; all other heights are above Mean High Water Springs.
Positions are referred to the WGS 84 compatible datum, European Terrestrial Reference System 1989 Datum (see SATELLITE–DERIVED POSITIONS note).
Navigational marks: IALA Maritime Buoyage System

There is one further rock symbol that I need to point out which, amazingly, is missing from the L'Aber-Wrac'h collection of rocky hazards. That is the symbol for a rock awash at Chart Datum. A rock, in other words, whose charted depth (or drying height) is zero. The symbol is a vertical cross with dots in each quadrant. These are not particularly common, although they do crop up from time to time – as in the approaches to the small port of Gorey on the east coast of Jersey (see illustration).

Three rocks off the east coast of Jersey that are awash at Chart Datum

Underwater Obstructions

There are quite a lot of man-made obstructions littering the seabed – some deliberate and some there by accident. Helpfully, though, the cartographer will try to indicate the position and nature of bottom obstructions to help you avoid fouling them and, in particular, to encourage you to anchor elsewhere. Many obstructions are marked with the abbreviation 'Obstn' and treated in precisely the same way as rocks or wrecks. They are coloured blue or green if appropriate to their depth, and they are given a necklace if they are likely to be dangerous to surface navigation.

10

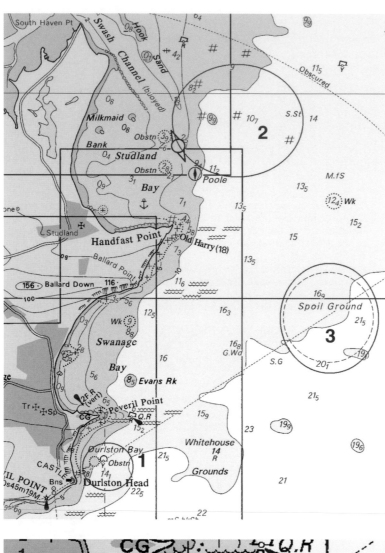

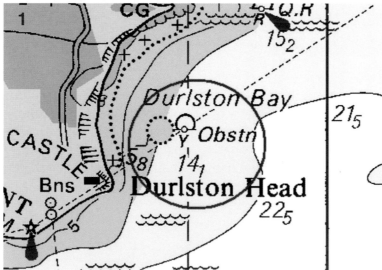

In this illustration, Circle 1 shows an unspecified bottom obstruction, less than 10 metres in depth (it is coloured blue), which is a danger to surface navigation, hence the necklace. It is marked by an unlit yellow special purpose buoy.

The four hash signs in Circle 2 represent foul areas which are not a hazard to surface navigation, for instance a chain on the seabed. This would not be a good place to anchor or trawl, but it's unlikely to cause you a difficulty on a routine passage.

And the **Spoil Ground** in Circle 3 may contain a variety of gremlins – probably the product of years of tipping waste and other debris onto the seabed. Once again, I would think long and hard before releasing my anchor in such an area. A more sinister note, which is still quite common around the coast, is 'Explosives Dumping Ground'. You have no idea what has been dropped overboard there, when, what state it's in or how badly it could damage you if it went off.

Depending on the geography, you will find many other kinds of obstruction around the world: fish traps, oil wells, pipelines, fish farms and simple piles or stakes. These are usually well-marked on the chart, often with an explanatory note, and you overlook them at your peril.

10

This illustration shows an oil production island off the coast of the UAE in the Persian Gulf. In Circle 1, you will see the square symbol of a production platform lit by a flare stack. Running out towards it from the island, and up to the north-west are a series of oil and gas pipelines – little tadpole symbols coloured magenta. It is interesting to note that these magenta symbols represent production pipelines, while the black symbol in Circle 2 represents intake or soil pipes.

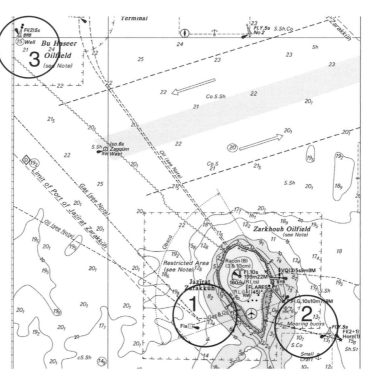

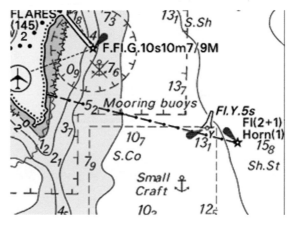

In Circle 3, there is a wellhead, which stands some 6 metres off the seabed, 15 metres below Chart Datum, and is considered a danger to surface navigation. For good measure (it is not far from the terminus of a traffic separation scheme), it is marked by an isolated danger buoy.

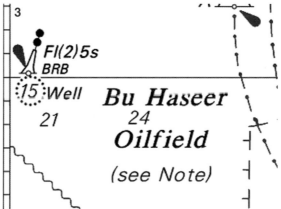

10

Deep-Water Obstructions

The entire seabed is strewn with cables of one form or another and these are generally charted where appropriate to the scale of the chart. Below, on a small-scale chart of the seabed between Orkney and Shetland to the north of Scotland, you can see a number of cables marked by corrugated magenta lines (see Circle 1).

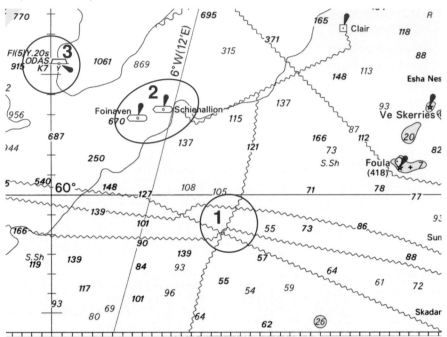

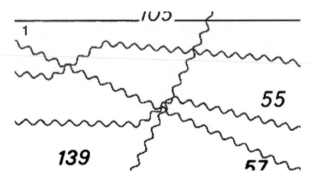

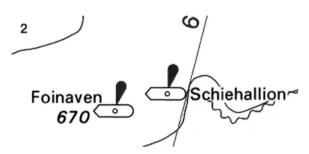

Additionally, two long sausage-like symbols are shown in the Circle 2. You don't see these very often, but this symbol represents a moored storage tanker – you would find the light characteristics on the larger-scale chart.

And in Circle 3 there is an **ODAS buoy**. ODAS stands for Ocean Data Acquisition System, and these buoys are placed at strategic intervals around the oceans to report on water and meteorological conditions. They are, among other things, crucial feeders for your daily weather forecast.

11 Navigation Restrictions and Limits

With growing levels of regulation, and exploitation of both the sea and the seabed, shipping is becoming increasingly restricted in coastal and some offshore areas. Before venturing into some of these busier waterways it is worth making sure that you understand the environment and its regulations as well as you can.

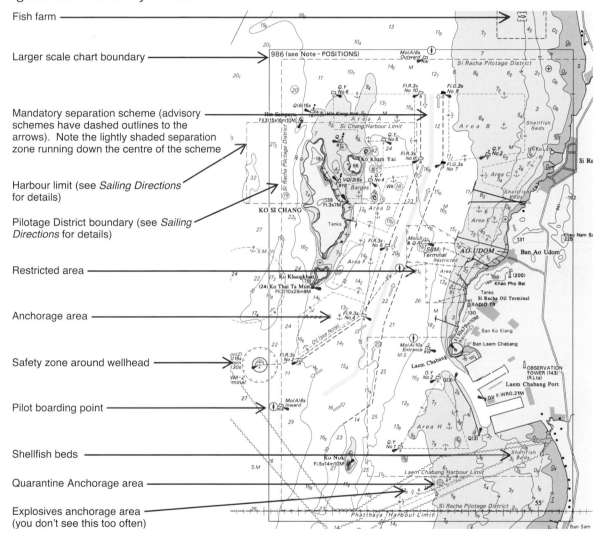

Fish farm

Larger scale chart boundary

Mandatory separation scheme (advisory schemes have dashed outlines to the arrows). Note the lightly shaded separation zone running down the centre of the scheme

Harbour limit (see *Sailing Directions* for details)

Pilotage District boundary (see *Sailing Directions* for details)

Restricted area

Anchorage area

Safety zone around wellhead

Pilot boarding point

Shellfish beds

Quarantine Anchorage area

Explosives anchorage area (you don't see this too often)

Welcome to Pattaya Harbour, on the coast of the Gulf of Siam, Thailand. This is clearly a busy port and, by the look of it, sensibly regulated too, with heavy commercial activity, fish farming and tourism all existing side by side. The price of this harmony is that shipping, and particularly visiting shipping, is much more constrained than in more open, less busy waterways. Even if you are going to take a pilot (and I imagine that it is compulsory in the area covered by this chart), you must still give the chart and the *Sailing Directions* a little time the evening before, to identify the constraints, the traffic flows and the various regulations that apply to you.

There is barely a magenta entry on this chart that is not of significance to almost every category of maritime traffic.

11

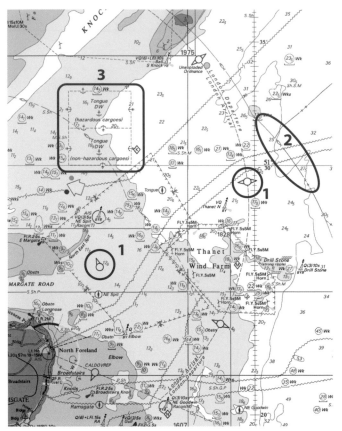

Next is a rather simpler chart, showing the mandatory reporting points for vessels entering and leaving the Thames. This is also a complex, busy piece of water, which covers the junction of the Straits of Dover Separation Scheme and the entrance to the Thames. It is, moreover, beset by shallow, unstable sandbanks, occasional tidal surges and strong tidal streams.

The duck-billed symbols for VHF reporting points are shown in Circle 1. This is the point at which you are expected to call the controlling VTS[1] when you are travelling in the direction of the beak.

Note too the national fishing limits (Circle 2), which endearingly are drawn as an intermittently dashed line with a small representation of a fish at intervals along it.

And finally, the Tongue deep-water anchorage, segregated between hazardous and non-hazardous cargoes, at Circle 3.

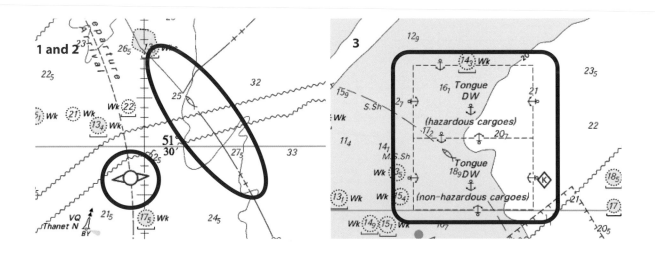

[1] VTS stands for Vessel Traffic Service, and it is the abbreviation given to the Control Room responsible for safety and management of a particular stretch of water. Some are more tightly controlled than others.

11

The Dover Strait

A few miles south from the mouth of the Thames lies the Dover Strait.

This is one of the busiest international seaways in the world, used by more than 400 large vessels each day. With shared jurisdiction between France and the United Kingdom, narrow shipping channels, a shallow, mobile seabed and strong tidal streams, there is little margin for error. It is for that reason that the regulations, and the charted depiction of this critical area of water, are kept as simple as possible.

From the point of view of cartography, areas of this operational intensity need to be portrayed as simply as possible, with clear symbology and unmistakeable

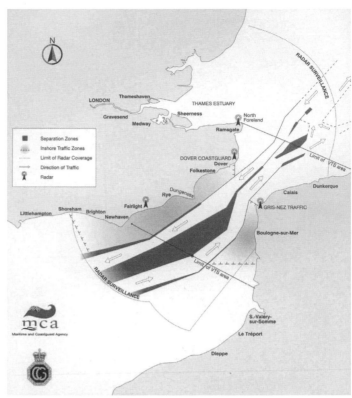

The Dover Strait Traffic Separation Scheme

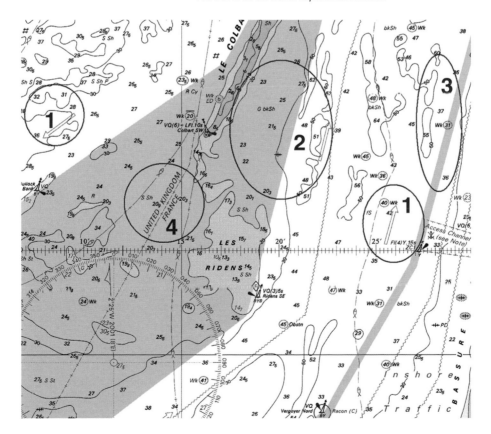

11

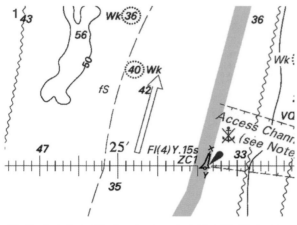

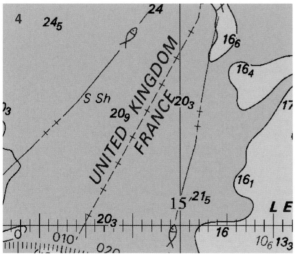

indicators of where a vessel may, and may not, venture. The chart-maker's problem is doubly complex here because the nature of the bottom is not simple and requires accurate, detailed charting.

So if you look at the chart of the Dover Strait, just south of Cap Gris Nez, you will find a pretty clear and uncluttered layout, with detailed bottom topography (many of the passing ships have only a few feet of clearance below their keel) and strongly shaded magenta boundaries to inform the passing traffic of the boundaries of the shipping lanes (Circles 2 and 3).

The direction of traffic flow is depicted by long, hollow arrows and repeated at intervals along the traffic lanes (Circle 1), so that there is absolutely no ambiguity about the purpose of the lanes.

And, because it matters in highly regulated international water space, the international boundary between British and French jurisdiction is shown with great clarity throughout the Straits (Circle 4).

And In Conclusion

There are any number of restrictions and limits shown on charts, ranging from seal sanctuaries to military restricted areas and from sensitive sea areas to seaplane operating areas. They are all painstakingly set out in Section N of *Chart 5011* (page 164 in this book). The point is that many of them will not welcome unscheduled visitors, and you really don't want to be straying into regulated areas unwittingly.

You must be particularly aware of international boundaries and the limits of territorial waters. In many of the more sensitive parts of the world, international borders are still disputed and the chart may not accurately represent the views of all parties. If the limit of territorial waters is in any way unclear, you must seek clear, unambiguous advice before going there. This is one more area where it would be foolhardy to rely unambiguously on the advice of your chart, which may be out of date or insufficiently clear on the various counter-claims of interested parties.

12 Radar and Radio

You are heading south through the Straits of Dover in heavy weather. You know that the Varne Bank, lying in the centre of your shipping lane, is marked by a single light buoy at its north end. How will you distinguish this particular buoy from any other one, or indeed from any slow-moving contact, on your radar?

The answer is that the Varne light buoy has a RACON mounted on it, which flashes the Morse code for 'T' (━) periodically, when interrogated by a passing radar.

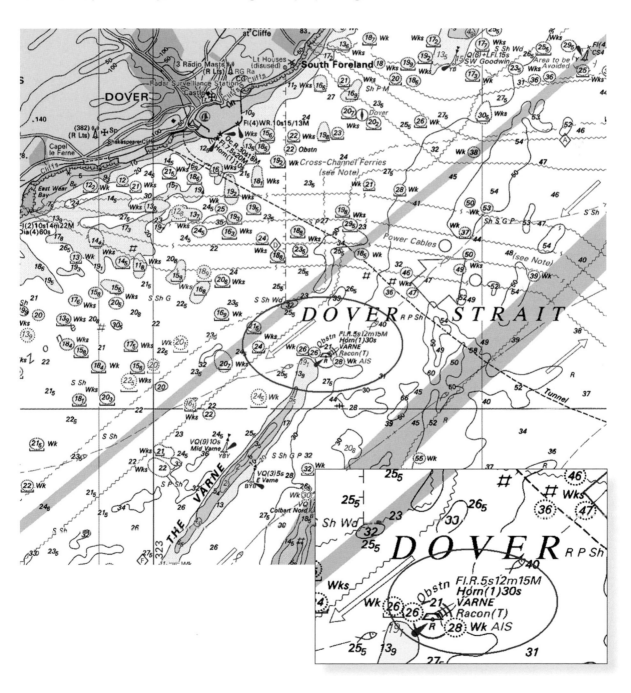

12

This long flash comes up behind the radar echo on the screen and is big and bold enough to be pretty much unmistakeable. The pulse of the Morse symbol is often 2–3 miles in length on the radar screen, and it paints only intermittently. RACONs may also be fitted to lighthouses or light vessels. Where they are fitted to buoys marking new dangers, such as recent wrecks in busy waters, they will generally be coded 'D' (━ ••). Self-evidently these 'new danger buoys' may not have been charted, but you will need to recognise them, and the sight of this Morse signal should also alert you to the presence of an unexpected problem.

You may also see from time to time a transponder marked on the chart as **Ramark**. These transmit continuously: they do not need to be triggered by a radar pulse, so they are not, precisely speaking, transponders – just radar beacons. The drawback with a Ramark is that, as a constant transmitter, it will display as a single radial line on the bearing of the transmitter, stretching from the centre of your screen to the edge. There is no way of determining the range of the transmitter, as there is with a RACON.

Name RC	Non-directional marine or aeromarine radiobeacon
RD 269·5° RD	Directional radiobeacon with bearing line
Lts≠ 270° RD 270° RD	Directional radiobeacon coincident with leading lights
RW	Rotating pattern radiobeacon
Consol	Consol beacon
RG	Radio direction-finding station
R	Coast radio station providing QTG service
Aero RC	Aeronautical radiobeacon
AIS	Automatic Identification System transmitter
AIS AIS	Automatic Identification System transmitters on floating marks (examples)

There are many other radio and radar beacons that will be displayed on your chart, and whose characteristics will be set out in detail in the relevant volume of the *Admiralty List of Radio Signals*. The common characteristic of their chart symbols is a small black dot with a magenta circle around it, amplified by a set of explanatory initials.

You may also come across a Search and Rescue Transponder (**SART**), the purpose of which is solely to highlight a vessel that is in distress. It produces a string of 12 short pulses on the radar screen, and should be unmistakeable from any other transponder signal.

13 Tides, Tidal Streams and Currents

At the risk of stating the obvious, there are three effects of moving water that are of interest to the mariner: height of tide, tidal stream and current. Height of tide obviously affects the under-keel clearance of the vessel, tidal stream is the horizontal movement of water due to the influence of heavenly bodies and a current is a more-or-less constant flow of water, normally over an extended distance.

All three are dealt with separately on a chart.

Height of Tide

We have already covered the fact that the charted depths are shown as the depth of water *below* Chart Datum (which is usually set at or around the level of lowest astronomical tide, or LAT), and the height of tide is the depth of water at any one time *above* Chart Datum. Accordingly, the total depth of water at that point is the sum of charted depth and height of tide.

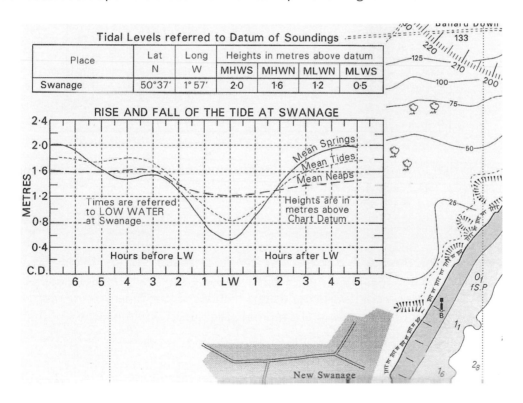

Any moderately competent mariner will be able to calculate height of tide using Standard and Secondary Port information. Some electronic aids to navigation can also do this but, unless you are absolutely certain of the source data that your grey box is using, I would not give up the tide tables, notepad and pencil.

Some places are a little more tricky for tidal calculations than others. Around the Western Solent and Poole Harbour, for instance, there tends to be a double high tide – so tidal predictions are, in this case, shown on the pilotage chart, referenced to low water (see diagram). This is not particularly complicated to work out, and it does give a more accurate prediction of tidal height.

Tidal Range

For any sort of precise navigation you need to understand the size of the tidal range: the difference in height between high and low water. By comparing the range on any given day to the range of spring and neap tides, you will be able to interpolate the strengths of tidal streams, etc. Most coastal charts set out the height of mean high and low water levels at springs and neaps in a table on the margins of the chart. They can also be found in the *Admiralty Tide Tables*.

In this example, taken from a chart of south-east England, you can see that:

The mean spring range at Eastbourne is: 7.4 – 0.7 m = 6.7 m
The mean neap range is: 5.4 – 2.1 m = 3.3 m

Tidal Levels referred to Datum of Soundings						
Place	Lat N	Long E	Heights in metres above datum			
			MHWS	MHWN	MLWN	MLWS
Newhaven	50°47′	0°04′	6·7	5·0	1·9	0·4
Eastbourne	50 46	0 17	7·4	5·4	2·1	0·7
Hastings	50 51	0 35	7·6	5·8	2·2	0·7
Dungeness	50 54	0 58	7·8	5·9	2·5	0·9
Dover	51 07	1 19	6·8	5·3	2·1	0·8
Boulogne–sur–Mer	50 44	1 35	8·8	7·2	2·6	1·1
Dieppe	49 56	1 05	9·3	7·4	2·5	0·8

For offshore data, see Co-Tidal Charts 5057 and 5058.

If you calculate that today's tidal range is 5.0 m, you are at exactly halfway between the springs and neaps ranges. This is known as '50% springs'.

Tidal Streams

In areas where the information is a little sparse, the tidal stream is often indicated by an arrow showing the direction and the rate of flow. These are the peak rates of the flood and ebb stream (they will almost always be pointing in opposite directions) at the mean-spring rate. You would clearly need to do a bit of interpolation to estimate the rate at different states of the tide, and different stages of the neaps / spring cycle.

It's worth reading the warning contained in the preface of Volume 1 of the *Admiralty Tide Tables*, which says:

'The tidal streams in European waters are, for the most part, of the same type as the tides. i.e. they are semi-diurnal in character *(twice daily)*. They can therefore be predicted by reference to a suitable Standard Port by tables printed on the published charts, and there is no necessity for daily predictions to be published. In some other parts of the world, however, the pattern of tidal stream is entirely unrelated to the pattern of the tides and in these cases daily predictions are necessary; such predictions will be found in Volumes 3 and 4 of the *Admiralty Tide Tables*.'

More accurate and thorough information on tidal streams can be found by using the tidal stream diamonds. These are the small diamond shapes that you will find scattered across a number of coastal charts in positions where a tidal stream recorder has been laid. They provide both direction and speed of the water flow at all stages of the tide.

Tidal Stream – mean spring rate

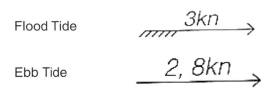

Flood Tide

Ebb Tide

13

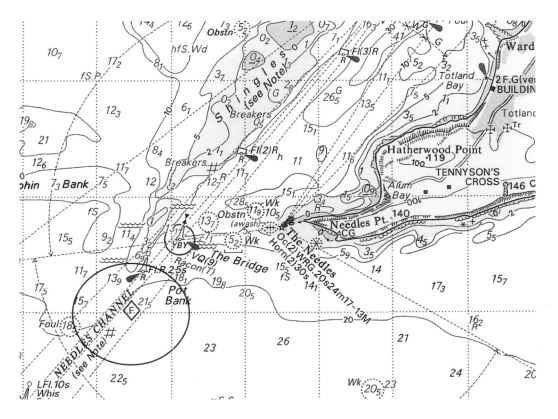

Here, off the Needles on the south coast of England, you will see tidal stream diamond **F** has been plotted in a position where the rate and direction of the tidal stream is of some importance to passing mariners, squeezed between the southern tip of the Shingles Bank and the off-lying shoals from the western tip of the Isle of Wight.

Such diamonds are referenced to a table which will be displayed somewhere in the margins of the chart. You need to find the table, identify the data relating to your specific diamond, both by letter and by position, and

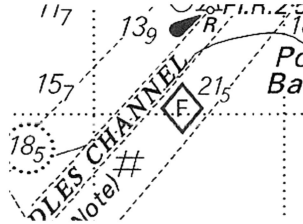

this gives you access to three columns of data. The first, which is printed in bold script, shows the direction in which the water is flowing (in degrees, true). The second and third show the mean spring rate and the mean neap rate, respectively.

13

If you know how today's tidal range compares to the spring rate, you can find, through a little judicious interpretation, the direction and rate of tide at any time, referenced to high water at the closest Standard Port, in this case Portsmouth.

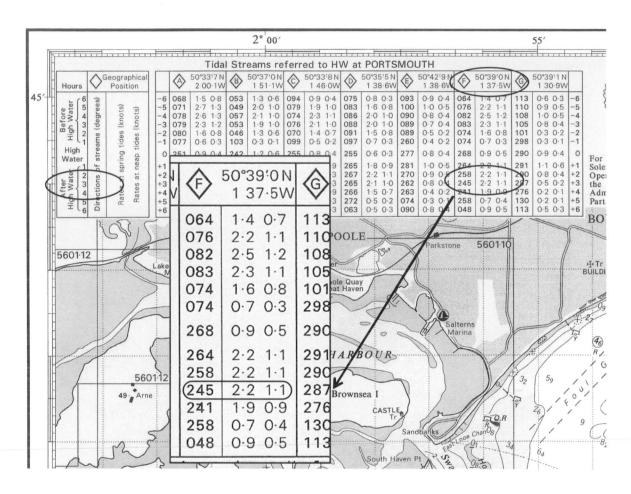

I have circled the data for HW + 3 hours: three hours after high water, where the flow is in a south-westerly direction (245°T) at 2.2 knots (springs) and 1.1 knots (neaps).

If, as in the Eastborne example[1], you are at '50% springs', the speed of the tidal stream at diamond F, three hours after high water, will be:

1.1 + 0.5 x (2.2 – 1.1) = 1.6 kts

In some places close to the United Kingdom, the UK Hydrographic Office has reproduced this data in a series of chartlets, called the *Tidal Stream Atlas*. This uses exactly the same data, but displays it in a geographical format, and the rates of spring and neap tides are shown alongside selected arrows in tenths of a knot.

---

[1] The rate depends on date, not geography. If it's 50% springs in Eastbourne, it will be 50% springs elsewhere too.

You will see from this extract of the *Solent Area Tidal Stream Atlas* that the arrow at the same position and time, three hours after high water, is shown as **11, 22** or 1.1 knots neap rate and 2.2 knots spring rate. Clearly, the direction of the arrow shows how the water is flowing.

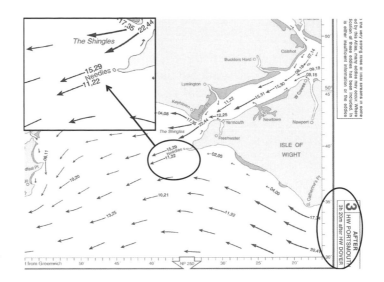

Tidal Coefficients

In some areas of the world, and the French coast springs to mind, you will find that the local hydrographer provides a table giving the tidal strength on any given tide as a 'tidal coefficient'. You do need to check the local convention, but the French coefficient tables run from 20 to 120, with 95 being the mean spring level and 45 the mean neap level. This provides a relatively simple tool for calculating, as a proportion of the spring rates, the tidal stream and height range on any given tide.

Ocean Currents

Ocean currents are found in deep water all round the world, and they are permanent or semi-permanent flows of water, largely in response to an enduring wind (like, for instance, the Trade Winds). Their rate and direction may be variable, and they are marked on small-scale charts with a sort of corrugated arrow, alongside which is written the expected rate of flow. It is important to compare this prediction with your own experience at the time: currents do vary and observation of their rates is often less

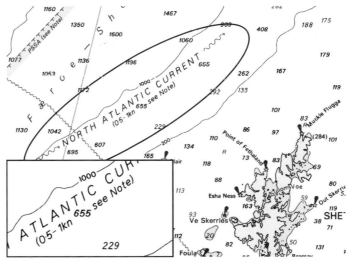

than systematic. A rule of thumb is that the rate of current flow is approximately 1/40th of the mean, sustained wind speed, clear of land and with a good fetch. That means that a 25–30 knot trade wind, blowing from the east, will set up a westerly set in the water of about 0.75 knots in the open ocean.

Sometimes, but not always, you will know that you are crossing the border of an ocean current because you will experience a rapid change in seawater temperature. This phenomenon is frequently associated with fog.

13

Turbulence

Coastal charts will often identify areas of likely turbulence in the form of overfalls or eddies. In many cases, you can work out likely areas of disturbed water for yourself: they are likely to occur where there is a strong tidal stream and an uneven bottom (e.g. the Pentland Firth); where in shallow water there is a brisk wind over a strong tide; where a point or peninsular causes two tidal flows to meet at an angle (e.g. Portland Bill); or where a long oceanic swell comes into shallow water and starts to break (e.g. the North Shore of Hawaii). In every case, without exception, the prudent mariner should check the likely severity of the turbulence in the *Sailing Directions*.

Overfalls, rip tide, races

Eddies

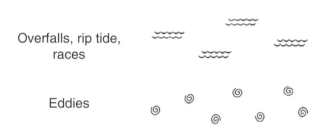

Helpfully, the hydrographer points out these likely areas of turbulence on the chart. It is worth taking note of these symbols, particularly if you are in a smallish vessel. The Weymouth RNLI lifeboat crew report a confused swell up to 4 metres in height which can occur in the Portland Bill race. Depending on the direction of wind, it may often be breaking too. This is not a comfortable place to be.

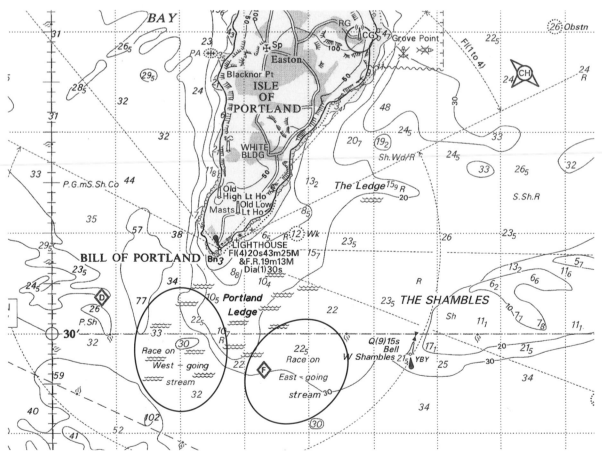

The Portland race is well-marked on the chart, and changes position according to the direction of the tidal stream.

14 Small Craft Features

I have never really been convinced of the need to over-print small-craft facilities on a mainstream navigational chart, although I recognise the commercial advantage for any chart publisher in catering for the leisure market. This information is more thoroughly covered in the *Sailing Directions* or small-craft pilot books.

In any case, on some of the Leisure Folio Charts, the Hydrographer has gone to considerable trouble to identify facilities which may be of value to small craft, and provide useful supporting information. Leisure facilities are all set out in Section U of *Chart 5011* (page 184 in this book). You may also be interested in NP109 – North West Europe catalogue, which contains a full list of Leisure Chart Folios, as well as standard nautical charts.

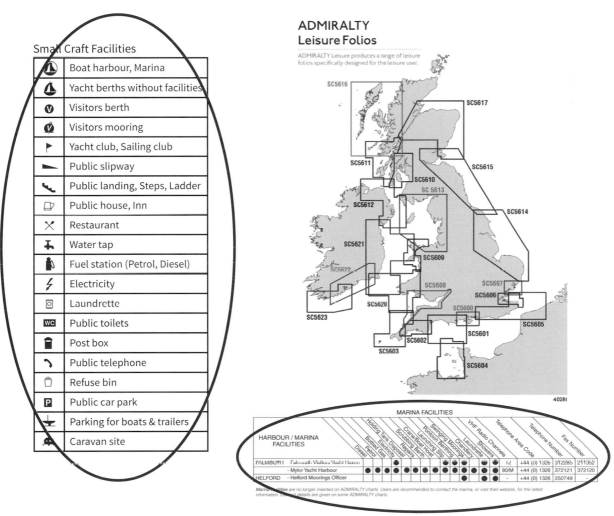

Small Craft Facilities

⚓	Boat harbour, Marina
⛵	Yacht berths without facilities
ⓥ	Visitors berth
ⓥ	Visitors mooring
⚑	Yacht club, Sailing club
◣	Public slipway
↖	Public landing, Steps, Ladder
⌂	Public house, Inn
✕	Restaurant
ⵔ	Water tap
⛽	Fuel station (Petrol, Diesel)
⚡	Electricity
⊚	Laundrette
WC	Public toilets
⬛	Post box
↰	Public telephone
⬛	Refuse bin
P	Public car park
⛢	Parking for boats & trailers
⛺	Caravan site

ADMIRALTY
Leisure Folios

ADMIRALTY Leisure produces a range of leisure folios specifically designed for the leisure user.

40281

MARINA FACILITIES

HARBOUR / MARINA FACILITIES	Diesel	Petrol	Bottled Gas	Holding Tank Disposal	Electricity	Regatta	Scrubbing Berth	Crane/Boat Hoist	Launching Slip	Pontoon Berthing	Swinging Moorings	Chandlery	Laundrette	Showers	VHF Radio Channels	Telephone Area Code	Telephone Number	Fax Number
FALMOUTH Falmouth Visitors Yacht Haven															12	+44 (0) 1326	312285	211352
- Mylor Yacht Harbour	●	●	●	●	●	●	●	●	●	●	●	●	●	●	80/M	+44 (0) 1326	372121	372120
HELFORD - Helford Moorings Officer											●		●		-	+44 (0) 1326	250749	-

Marina facilities are no longer inserted on ADMIRALTY charts. Users are recommended to contact the marina, or visit their website, for the latest information. Limited details are given on some ADMIRALTY charts.

On the chart overleaf of Hugh Town, St Mary's, two fuel berths, two WCs, a public telephone and, on Garrison Hill, a campsite are shown, together with a number of more conventional symbols like those for a hospital and a post office.

14

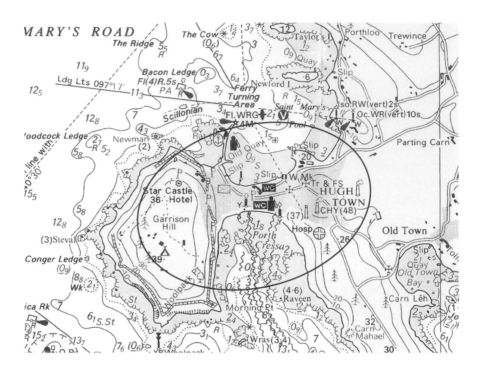

Coming a little further east to the mouth of Weymouth harbour, there are a number of interesting chart symbols on the waterfront – not all of them specifically designed for small vessels – which are worth a mention.

In Circle 1, the 'McDonald's Quarter Pounder' symbol represents the Customs Office. Circle 2 is the berth number in the commercial port: in this case Berth 4. Circle 3 is the lifeboat mooring. And circle 4 has a number of interesting symbols in a small part of the Quay. Starting from the south western end, they are the Coastguard office, a yacht club (the Royal Dorset), Harbourmaster's office and the symbol for a fishing port.

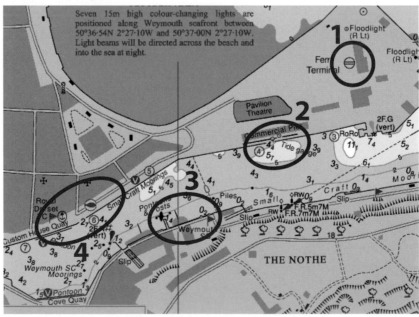

15 Other Charting Authorities

Other UK Charts

The UK Hydrographic Office (UKHO) is the only authority in the United Kingdom which has the resources and the international network to provide a virtually complete global charting service with 3 300 paper navigational charts, worldwide digital charts and 220 publications available through the *Admiralty Chart Catalogue*.

There are a number of other commercially produced charts available in the United Kingdom, many of which are specifically intended for use in small craft. These charts use data that is made available by the UKHO and other responsible authorities, and they issue updates regularly. There is no reason to assume that they will not be as accurate and reliable for small craft as a UKHO chart, and they may well be easier to use. However, unless they say otherwise, I would be unwilling to use these small-craft charts for larger vessels, which may find themselves restricted by the limited range of charts and the absence of important risk-based features, like the Source Data diagram.

Imray Charts carry a large range of charts for yachtsmen, for small commercial vessels and fishing vessels under 24 metres.[1] The source data for these charts is drawn from the UKHO or other relevant hydrographic offices, port authorities, Trinity House, etc. Imray has an extensive geographical coverage, which includes the United Kingdom and the coast of Europe from Denmark to Gibraltar, the Mediterranean, Atlantic and Caribbean. They are printed on water-resistant paper and have a different colour scheme from Admiralty charts, but many of the symbols are either common or similar. Corrections to Imray charts can be obtained at www.imray.com, or direct from the publishers.

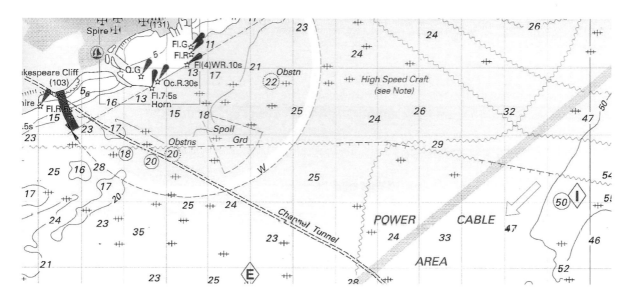

---

[1]Imray charts have been awarded official recognition by the MCA for carriage on Code vessels under 24 metres and fishing vessels under 24 metres in length on the basis that quality control is formally monitored at all stages of production, that charts are easily correctible and that source data is derived from official hydrographic information.

15

If you are navigating a small craft, the choice of chart supplier will be very much a matter of taste, trust and familiarity. Properly maintained, any one of these brands should provide you with sufficiently accurate and recent information to navigate safely.

Rest of the World

Many countries that have a coastline publish their own charts. The bigger and wealthier maritime states have the resources to run a global hydrographic office and collate the information provided by other countries into their own chart portfolios. There is little difference between the quality and detail of charts produced in the most rigorous countries, particularly for those charts that have adopted the common symbology of the International Series.

It would be invidious to say that one country's charts were 'better' than others. However, standards and thoroughness do vary around the world, and some hydrographic offices may not be as thorough in checking and vetting the incoming data as others.

US Charts

Charts of US waters are produced by the National Ocean Service (NOS), a part of the National Oceanographic and Atmospheric Administration (NOAA). US charts are numbered systematically, with the first of the five digits representing the chart region, the second identifying the sub-region and the last three being the number of the individual chart within that geographical area.

American charts (or a chart published by a European agency which has been 'adopted' from the United States) will generally show soundings in fathoms and feet, but you should always, as a matter of course, check the information in the chart Title Block. US charts use national symbols and conventions, although *US Chart No 1*, the chart symbol handbook, is very similar to the UK version, *Chart 5011*. US charts are not always orientated on north, and they use different vertical datums: Chart Datum is often referenced against mean lower low water[2] (MLLW) instead of lowest astronomical tide (LAT).

Finally, and this is just for the cognoscenti, NOAA uses the North American Datum 1983 (NAD83) instead of WGS84, although the two are 'navigationally equivalent', to quote the NOAA website. None of this should cause a competent navigator any difficulty; it is merely a matter of understanding the data that you are presented with and, as always, taking time to study the chart before you use it.

NOAA also publishes a suite of navigational publications which are very similar to those published by the UKHO.

---

[2]On any given day, there are two low waters. MLLW records the lower of these two heights and averages them over a 19-year period.

16 GPS

Gone are the days when mariners would feel content when they completed an ocean passage and arrived at their landfall within 10 nautical miles of accuracy. And gone too are the navigators like me who would have considered such a landfall a most remarkable fluke (although an important part of navigation training was learning to maintain your composure in the face of either happy coincidence or sheer desperation). Today, with GPS sitting on every bridge, in every mobile phone, car, wristwatch and in fact in anything that moves, navigators expect to have pinpoint accuracy at any time, in any place and in all conditions.

The danger, of course, lies in relating the position on your GPS to your position on the earth. Inside its microscopic memory, the GPS system uses a model[1] of the planet's shape to calculate a position from the various satellite radio signals. Your chart-maker, meanwhile, uses his own model of the earth's shape to allocate positions to the chart that he has drawn. A lot of care is put in to ensure that both the satellite and the cartographer are using the same model. If they aren't (and there are still a lot of published charts around that are on a different datum from the GPS system), you may find that any GPS position plotted on the chart is substantially in error.

The model used by the US GPS system, which the great majority of us take as standard, is called WGS84, or the 1984 World Geoditic Survey. If your chart is plotted to the WGS84 standard, any position taken from the US GPS system can be plotted on it without correction.

By way of an example, look at the position of South Foreland Light, on the south coast of the United Kingdom, when plotted according to various different horizontal datums:[2]

Geographical Position	Horizontal Datum
51° 08'.39N 1° 22'.37E	Referred to OSGB36 Datum (the local datum for Great Britain)
51° 08'.48N 1° 22'.35E	Referred to European (1950) Datum ED50 (the continental datum)
51° 08'.42N 1° 22'.26E	Referred to WGS72 Datum (the obsolete worldwide datum)
51° 08'.42N 1° 22'.27E	Referred to WGS84 Datum (the worldwide datum used by GPS)

Both OSGB36 and ED50 are about 135 metres from the position of WGS84.

A large part of the world's charts have yet to be converted to the WGS84 datum. So, if you are going to use GPS, always check the chart's horizontal datum, which will be announced either in the margin or in the Title Block. If it is drawn to WGS84, all well and good. If it is drawn to another datum, make sure that you religiously apply the corrections that you will find in the chart's Title Block.

---

[1]These models are known as 'horizontal datums'.
[2]Taken from the *Admiralty List of Radio Signals*, Volume 2.

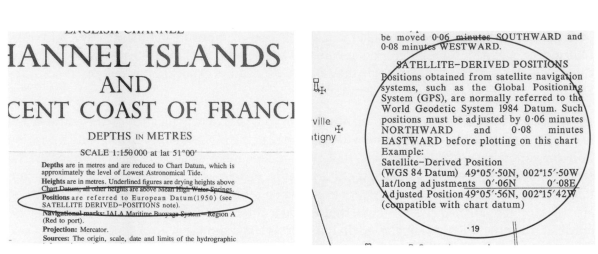

Accuracy of GPS

Let's just assume that you apply all of the proper datum-derived corrections. Are you now safe to use GPS as a categorical position reference? The answer is a sort of equivocal 'Yes, just about.'

In May 2000, the US Department of Defense, which operates the GPS system, removed a sort of artificial wobble that it had applied to the system which gave uncorrected signals a significant and random error. This had been applied for reasons of national security, and was called 'selective availability'. This has now gone, and the advertised accuracy of the system is plus or minus 7.8 m (25.6 ft), with 95% probability (GPS.com). But don't rely on this sort of accuracy; random obstructions, the number and orientation of the satellites, and unspecified electronic gremlins can all degrade the accuracy of your little grey box. It is accurate, but the wise mariner will still treat it with a healthy dollop of suspicion too.

In my mind, this is a real problem: we all start to believe that GPS is super-accurate and totally infallible. I can still show you the scars in the front of my boat that I sustained hitting a buoy by navigating out of a French port at dead of night with no moon, relying solely on my GPS and chart plotter. I should have known better: it could so easily have been a rock rather than a buoy, but I had been lulled into the false security of believing that the chart plotter was telling me the truth. It wasn't. It's always a good idea to periodically verify your GPS position – by taking a bearing, or a sounding and making sure that it agrees. On a digital chart, you might want to turn on the radar overlay from time to time to check that the coastline is where it ought to be.

But the most important thing about using GPS is that none of us should ever – ever – forget that somewhere in the middle of the United States of America, there is a big switch with 'ON – OFF' written alongside it, and the man or woman who operates that switch doesn't need your permission to switch it over to the 'OFF' position, at which point you may as well use your GPS set for decoration

– it won't be much use for anything else. One of the great lessons of seamanship is to trust no one, and always have a back-up system available.

Differential GPS

Differential GPS (DGPS) is one way of reducing the errors on GPS still further, and it is often used for coastal survey work. In effect, you set up a GPS receiver on the dry land with a VHF transmitter alongside it. Since you know what the precise position of the GPS receiver is, you can compare the GPS received position with the known position and work out the error at any one time.

That error is sent out through the VHF aerial so that a suitably equipped vessel within earshot can pick up the VHF signal and remove the error from its GPS set. DGPS works over a limited range only and you really don't need this sort of accuracy for anything but specialised applications or, occasionally, for bringing large vessels into port. Details of DGPS systems are set out in the *Admiralty List of Radio Signals* Volume 2.

16

17 Digital Charting

Digital charts are here to stay. They are accurate and less bulky to store than paper charts; they can be updated in an instant, despatched electronically to a ship in the middle of the ocean and they can carry a huge amount of supplementary data. An electronic chart plotter, moreover, will provide you with a real-time display of the ship's position on the chart, and a superimposed radar and AIS picture can give you the sort of traffic awareness that our predecessors could only dream of.

An updated digital chart from a reputable supplier is every bit as accurate as its paper equivalent but, since it draws its data from precisely the same sources, it has precisely the same potential for inaccuracy as a paper chart. A prudent mariner should always, therefore, be aware of the accuracy of the source data when using a digital chart, and he must of course keep his chart folio, whether paper or digital, properly corrected.

A Few Abbreviations Relating to Electronic Charting

ECDIS.   Electronic Chart Display and Information System. This term refers to a navigation information system, with adequate back-up arrangements, which has been accepted by the IMO as complying with the up-to-date charting requirements in accordance with Chapter V of the 1974 SOLAS Convention.

ECS.   Electronic Chart System. ECS is a navigation information system that electronically displays vessel position and relevant nautical chart data and information from an ECS Database on a display screen, but does not meet all the IMO requirements for ECDIS and is not intended to satisfy the SOLAS Chapter V requirements to carry a navigational chart.

ENC.   Electronic Navigation Chart. These are vector electronic charts (see below) that conform to International Hydrographic Office specifications.

RNC.   Raster Navigation Chart. These are raster electronic charts (see below) that conform to International Hydrographic Office specifications.

Electronic Chart Display and Information Systems

ECDIS is the name given to electronic navigation systems which plot a real-time position of the ship on a digital chart. These can vary from hand-held GPS systems to substantial fixed consoles, on the bridge and elsewhere, interfaced with ship systems such as radar, AIS, log, gyro, etc.

It is important to understand how reliable your electronic chart system is: digital charting is increasingly being adopted as the principal means of marine navigation but, if the mariner is really going to rely on an electronic system, he or she must be confident that it has the integrity and the back-up to keep going through thick and thin. For that reason, the IMO has created two categories of chart display systems. The first is an ECDIS, which meets IMO/SOLAS carriage requirements (in other words it is reliable enough to act as a stand-alone system). The second is an ECS, which does not meet IMO/SOLAS requirements. ECS can only be used as a navigation aid, and a full complement of updated paper charts must be both carried and used for navigation.

Mandation

The 2011 amendment to Chapter V of the SOLAS Convention specifies that:

> All ships, irrespective of size, shall have nautical charts and nautical publications to plan and display the ship's route and intended voyage and to plot and monitor positions throughout the voyage. An electronic chart display and information system (ECDIS) is also accepted as meeting the chart carriage requirements...

It goes on to say that just about all large ships 'engaged on international voyages' should by now (2017) have been fitted with an ECDIS system.

Sensibly, this amendment also talks about the need to carry a back-up systems which, depending on approval of the flag state, could include:

- An appropriate folio of corrected paper nautical charts.
- An ECDIS system using ENCs with an independent power supply.
- A Raster Chart Display System using RNCs or ENCs, with an independent power supply.

It's important to recognise that moving from a paper system to a full ECDIS system is not just a matter of installing new hardware and downloading a few digital charts. It does require the users to be properly trained if they are going to use the system safely. The techniques of digital charting are different, and the symbols used on ENC are also different. These are covered in an Admiralty publication called *NP5012. Guide to ENC Symbols used in ECDIS*.

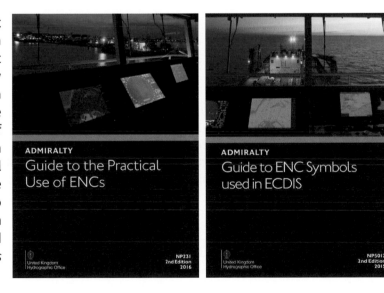

By the same token, a further publication called *NP231 Admiralty Guide to the Use of ENCs* is an important handbook for anyone who is using ECDIS at sea.

Digital Charts

Digital charting has evolved over recent years, breeding 2 distinct species of chart, called 'raster' and 'vector'. They have evolved in different ways, and both now look and behave differently, even though they both derive information from the same high-integrity sources.

17

Raster Charts

Raster charts are direct digital reproductions of the paper charts, made for display on an electronic chart plotter. As a result, all of the familiar symbology is there, just as it would be on a paper chart; the titles and source diagrams are available and anyone used to paper charts would find a raster display reassuringly familiar. The UK Hydrographic Office currently produces over 3200 raster charts on 11 CD-ROMs under the brand of ARCS (Admiralty Raster Chart Service). Compared to vector charts, raster information uses a large amount of memory; the chart picture consisted of thousands of small pixels, and there is no flexibility to add or remove information selectively. Furthermore, it is possible that horizontal datum and chart projections will change as you move from the area covered by one chart to the next (in exactly the same way as a paper chart can change. The shift required to display an accurate WGS84 position of the chart is contained in the metadata of the chart, and this can be applied automatically by your chart display system.

Raster charts are updated weekly by CDs issued from the UKHO, or online using the UKHO updating service.

With state approval, raster navigation charts (or RNCs as they are called) may be used on an ECDIS system, but when ECDIS is employed in this way, the IMO requires the system to be used in conjunction with an appropriate folio of up-to-date paper charts.

Increasingly, however, vector-based electronic navigation charts (ENCs) are being used in preference to raster charts as the preferred digital charting system in order to meet the requirements of the SOLAS Convention.

Vector Charts

Vector charts are now becoming the mainstay of digital charting, both for leisure and professional mariners. The reason for this is that the vector format is so flexible. Vector charts are not made up of pixels, but a mass of interconnected short lines ('vectors') that join up to provide the outline of various features on the chart. You can see the vectors quite clearly when you zoom in to the highest magnification; the features look crudely drawn and a little spiky. Since this system cannot easily reproduce detailed chart features, additional information is applied to the basic topographical background by means of a number of chart 'layers', so that the position of buoys for instance, or sounding data, the radar display, AIS, light characteristics, the height of tide and much more information can all be selectively added – or 'decluttered' – depending on your needs at the time.

This system of superimposed layers gives vector charts great flexibility and widespread appeal. Moreover, since the vector charting facility uses so much less memory than raster charts, a great deal of additional information can be added without substantial cost.

Vector charts that can be used on an ECDIS system are called electronic navigation charts, or ENCs. ENCs are seamless, with no boundaries between charts, and they are all drawn up using the WGS84 datum. They are updated weekly by CD, DVD or online.

Every week, the UKHO also publishes an ENC 'Information Overlay', which contains all Admiralty Temporary & Preliminary Notices to Mariners (T&P NMs) and highlights navigationally significant differences between ENCs and Admiralty paper charts. Note the use of 'navigationally significant':

there are important adjustments to the data provided on an ENC that ECDIS users would be well advised to take into account when planning a passage.

Source Data in Vector Charts

17

One important layer that is provided in UKHO ENCs, and may not appear in other, less rigorously managed, vector systems, is the CATZOC[1] source information layer.

Like any layer, this can be switched on or off as required, and it gives a graphical representation of the reliability and integrity of localised source data, ranging from six stars to 'unassessed', based on three factors: position accuracy, depth accuracy and seafloor coverage. One or other of these 'starry lozenge' symbols will be overlaid on any given part of the chart, providing you with an instant assessment of the reliability of the chart data.

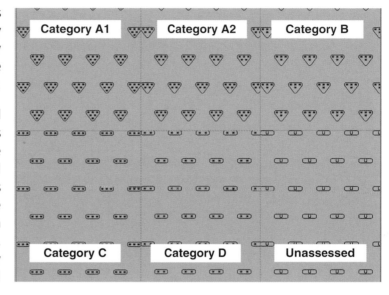

Here you will see an ENC screenshot, with the CATZOC layer applied, showing that this part of the coast of south-east England is surveyed to cat B standard.

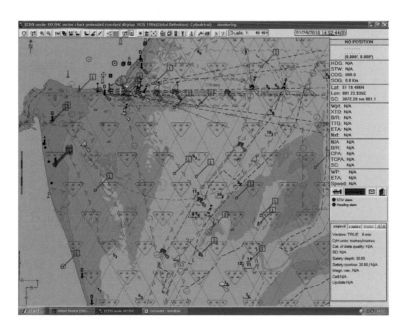

[1]CATZOC: a wonderful acronym, which stands for 'Category Zone of Confidence'.

17

The accuracy criteria in each zone are set out in Chapter 2 of *The Mariner's Handbook*, which reads as follows:

ZOC	Position Accuracy (metres)	Depth Accuracy (metres)	Seafloor Coverage	Stars on CATZOC overlay
A1	±5	$0.5\,m + 1\%$ depth^2	Full seafloor coverage. All significant features detected and depths measured.	6*
A2	±20	$1.0\,m + 2\%$ depth	Full seafloor coverage. All significant features detected and depths measured.	5*
B	±50	$1.0\,m + 2\%$ depth	Full seafloor coverage not achieved. Uncharted features, dangerous to surface navigation not expected, but may exist.	4*
C	±500	$2.0\,m + 5\%$ depth	Full seafloor coverage not achieved. Depth anomalies may be expected.	3*
D	Worse than Cat C	Worse than Cat C	Full seafloor coverage not achieved. Large depth anomalies may be expected.	2*
U	Unassessed			U

CATZOC is quite a useful tool. So much so that some charting authorities, like the UKHO, are now using it on their paper charts.

Digital Charts for the Leisure User

Many yachts and smaller boats are now equipped with chart plotters which use a variety of digital charting products. As a yacht owner, I have used Navionics products with considerable confidence over the years, and have not been disappointed by the quality or the presentation of the information. Indeed, for many small craft users, particularly when navigating without a large crew, digital charting provides a much more accessible way of determining your position and ensuring the safety of your passage. My hot tips for anyone using digital charts in a private capacity are as follows:

■ Always use a high quality supplier.

■ Keep your charts regularly updated. It will cost a bit, but it is worth it – honestly.

- Check your position with an external fix from time to time. Whether that is switching on the radar overlay to check that the coastline is where you expected it to be, or noting when a landmark passes through a cardinal point, and cross-checking with the chart. It all adds to your confidence that you are using reliable information.

- Don't blindly trust collision avoidance information derived from your chart plotter. Do what mariners are meant to do, and check obstructions and passing vessels visually for bearing movement.

- Always be aware of which layers of information that are switched on, and off. Decluttering is great – right up to the point where you hit a hazard that is displayed on one of the deselected layers.

- And always have a back-up system of navigation available. On a small boat, I would recommend having a sufficiently comprehensive folio of paper charts available (and updated) so that you can confidently navigate home in the event that a marlinspike goes through the chart plotter.

17

Symbols and Abbreviations used on Admiralty Paper charts

SYMBOLS and ABBREVIATIONS used on ADMIRALTY Paper Charts

CONTENTS

INTRODUCTION

General — NP5011 is primarily a key to symbols and abbreviations used on ADMIRALTY and International paper and raster charts and leisure folios compiled by the United Kingdom Hydrographic Office (UKHO). Variations may occur on charts adopted into the ADMIRALTY Series that were originally produced by another hydrographic office. Where these symbols and abbreviations are easily understood they will not be included as examples in this publication. Symbols and abbreviations shown on Electronic Chart Display and Information Systems (ECDIS) may differ from those described in this document; a key to such symbols is available: NP5012.

Schematic Layout of NP5011 — This edition of NP5011 is based on the "Chart Specifications of the IHO" (International Hydrographic Organization) adopted in 1982, with later additions and updates. The layout and numbering accords with the official IHO version of INT 1 (English version produced by Germany).

① ② Tracks, Routes **M**

④	④	
Tracks Marked by Lights → P	Leading Beacons → Q	③—Tracks

⑤ 1	2 Bns ⧧ 270·5° / 2 Bns ⧧ 270·5°	Leading line (the firm line is the track to be followed)	† Bn Bn *Bns in Line* 270°30' / *Ldg Bns* 270·5° / 270·5°	433.1 433.2 433.3

⑥ ⑦ ⑧ ⑨

① Section.

② Section designation. (In some nautical publications, this reference is pre-fixed "I", for International.)

③ Sub-section.

④ Cross-reference to terms in other sections.

⑤ Column 1: Numbering following the International "Chart Specifications of the IHO". A letter in this column, e.g. **a**, indicates a supplementary national symbol for which there is no International equivalent.

⑥ Column 2: International (INT) symbols used on ADMIRALTY paper charts. True to scale representations are to the left of symbols, where both are shown.

⑦ Column 3: Term and explanation in English.

⑧ Column 4: Other symbol or abbreviation used on ADMIRALTY paper charts, if different from Column 2. May be obsolescent or non-International.

⑨ Column 5: Not navigationally significant. Cross references to the "Chart Specifications of the IHO", S-4 (Part B, unless a reference letter to another part is given).

The mark † indicates that this representation or usage is obsolescent.
The mark # in Columns 2, 3 and 4 indicates that this symbol will only be found on paper charts adopted into the ADMIRALTY chart series. However, users should note that on such charts additional or different symbols not listed in this publication may be used. Where not easily understood, such symbols will be explained on those charts.

Metric Charts & Fathoms Charts — Metric units are introduced on ADMIRALTY charts as they are modernised (except for charts of the waters around the United States of America, where fathoms or feet continue to be used). Fathom and/or feet charts can be distinguished from metric charts by the use of grey for land areas, a note in the title block and in some cases by a prominent legend in the margin.

Chart Datum — On metric charts, the reference level for soundings is given under the chart title. On fathoms charts, the reference level for soundings may be given under the title; if not, it can be deduced from the tidal information panel.

Depths — The units used are given under the title of the chart. The position of a sounding is the centre of the area covered by the figures.

On metric charts, depths of less than 21m are generally expressed in metres and decimetres. Where source information is sufficiently precise, depths between 21m and 31m may be given in half-metres. All other depths are rounded down to whole metres.

On fathom charts, depths are generally expressed in fathoms and feet where less than 11 fms, and in fathoms elsewhere. Where source information is sufficiently precise, depths between 11 and 15 fms may be given in fathoms and feet. Older charts may show fractions of fathoms in depths of 10 fathoms or less, and some large-scale charts show all depths in feet.

On adopted or co-produced charts these ranges may vary.

Drying heights — Underlined figures on rocks and banks which uncover indicate heights above chart datum. They are given in metres and decimetres or in feet as appropriate.

Heights	Heights are given in metres or in feet above the charted height datum; details are given in the Explanatory Notes under the chart title. The position of a height is normally that of the dot alongside it, thus ·79. Parentheses are used when the figure expressing height is set apart from the object (e.g. when showing the height of a small islet). Clearance heights may be referred to a higher datum than other heights. In such cases this will be stated in the Explanatory Notes.
Bearings	Bearings are given from seaward and refer to the true compass.
Names	Names on ADMIRALTY charts are spelt in accordance with the principles and systems approved by the Permanent Committee on Geographical Names for British Official Use.

A second name may be given, usually in parentheses, in the following circumstances:
    a. if the retention of a superseded rendering will facilitate cross-reference to related publications;
    b. if, in the case of a name that has changed radically, the retention of the former one will aid recognition;
    c. if it is decided to retain an English conventional name in addition to the present official rendering.

Chart Catalogues	Details of ADMIRALTY charts are given in the "Catalogue of ADMIRALTY Charts and Publications" (NP131), and regional catalogue "North West Europe" (NP109), both published annually, and in the ADMIRALTY online Catalogue.
The Mariner's Handbook and other Publications	The Mariner's Handbook (NP100) includes information on the following:

The use of charts and the degree of reliance that may be placed on them; chart supply and updating; depth and height datums; names; related publications; navigation (including regulations, routeing, hazards and aids to navigation); tides and currents; general marine meteorology. A glossary of terms used on ADMIRALTY charts is also given.

Information about features represented on charts can also be found in the following publications or their digital equivalents:

ADMIRALTY Sailing Directions; ADMIRALTY List of Lights and Fog Signals; ADMIRALTY Tide Tables and Tidal Stream Atlases; ADMIRALTY List of Radio Signals; Annual Notices to Mariners; IALA Maritime Buoyage System.

How to Keep Your ADMIRALTY Products Up-to-Date	How to Keep Your ADMIRALTY Products Up-to-Date (NP294) provides comprehensive guidance on how to update both paper and digital ADMIRALTY charts and publications.
Copyright	ADMIRALTY charts and publications (including this one) are protected by Crown Copyright. They are derived from Crown Copyright information and from copyright information published by other organisations. They may not be reproduced in any material form (including photocopying or storing by electronic means) without prior permission of the copyright owners, which may be sought by applying, in the first instance, to the Copyright Manager, The United Kingdom Hydrographic Office, Admiralty Way, Taunton, Somerset TA1 2DN, UK.

Schematic Layout of an ADMIRALTY INT chart (reduced in size)

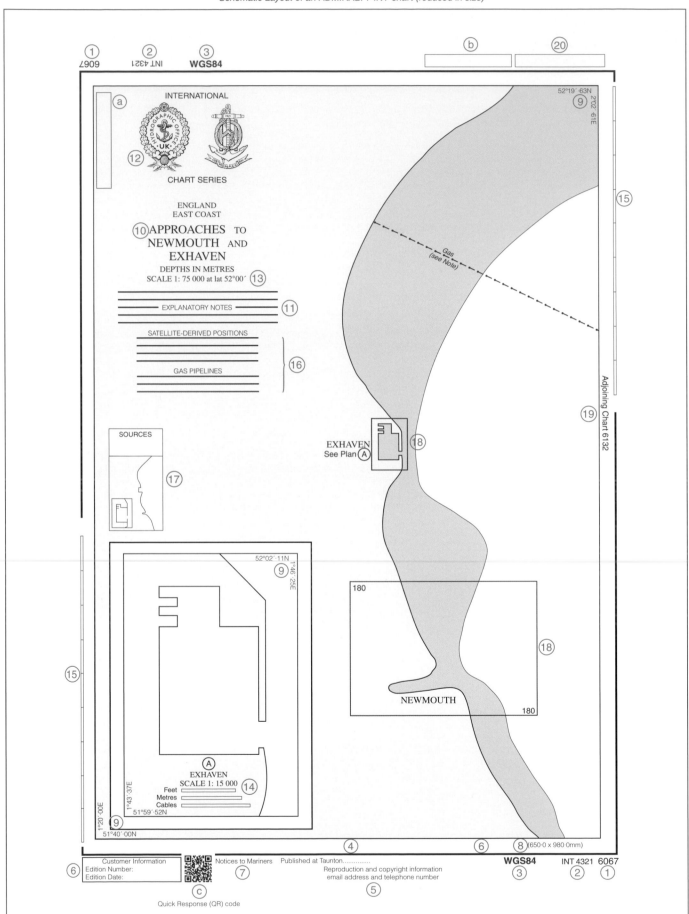

Chart Number, Title, Marginal Notes **A**

Magnetic Features ⟶ B   *Tidal Data* ⟶ H   *Satellite Navigation Systems* ⟶ S

(1)	*Chart number in the ADMIRALTY series.*	251
(2)	*Chart number in the International (INT) Chart series.*	251.1
(3)	*Use of WGS84 geodetic reference system. A reference to the depth units may be shown.*	201 255.3
(4)	*Publication note (imprint) showing the date of publication as a New Chart.*	252.1 252.4
(5)	*Reproduction and Copyright acknowledgement note. All ADMIRALTY charts are subject to Crown Copyright restrictions.*	253
(6)	*Customer Information: Edition Number and Date. (Charts revised prior to May 2000 have New Edition date at bottom right of chart)*	252.2
(7)	*Notices to Mariners: (a) the year dates and numbers of Notices to Mariners and (b) the dates (usually bracketed) of minor updates included in reprints but not formally promulgated (abandoned as a method of updating in 1986). (Charts revised prior to May 2000 have the legend 'Small corrections').*	252.3
(8)	*Dimensions of the inner neat-lines of the chart border. In the case of charts on Transverse Mercator and Gnomonic projections, dimensions may be quoted for all borders of the chart which differ. Some Fathoms charts show the dimensions in inches e.g. (38.40 x 25.40).*	222.3 222.4
(9)	*Corner co-ordinates.*	214
(10)	*Chart title. This should be quoted, in addition to the chart number, when ordering a chart.*	241.3
(11)	*Explanatory notes on chart content;* **to be read before using the chart**.	242
(12)	*Seals. Where an ADMIRALTY chart is in the International Chart series, the seal of the International Hydrographic Organization (IHO) is shown in addition to the national seal. Reproductions of international charts of other nations (facsimile) have the seals of the original producer (left), publisher (centre) and the IHO (right). Reproductions of other charts have the seals of original producer (left) and publisher (right); charts which are co-productions carry the seals of the nations involved in their production.*	241.1 241.2
(13)	*Scale of chart; on Mercator projection, at a stated latitude.*	211 241.4
(14)	*Linear scales on large-scale plan.*	221
(15)	*Linear border scales (metres). On smaller scale charts, the latitude border should be used to measure Sea Miles and Cables.*	221.1
(16)	*Cautionary notes (if any) on charted detail;* **to be read before using the chart**.	242
(17)	*Source Diagram (if any). If a Source Diagram is not shown, details of the sources used in the compilation of the chart are given in the explanatory notes (see 11).* **The Source Diagram or notes should be studied carefully before using the chart in order to assess the reliability of the sources**.	290-298
(18)	*Reference to a larger scale chart or plan (with reference letter if multiple plans on same chart).*	254
(19)	*Reference to an adjoining chart of similar scale.*	254
(20)	*Note 'IMPORTANT - THE USE OF ADMIRALTY CHARTS'. References to other publications.*	243
(a)	*Conversion scales. To allow approximate conversions between metric and fathoms and feet units. On older charts, conversion tables are given instead.*	280
(b)	*Copyright Notice.*	
(c)	*Quick Response (QR) code.*	

Magnetic Features ⟶ B   *Tidal Data* ⟶ H   *Satellite Navigation Systems* ⟶ S

B Positions, Distances, Directions, Compass

	Geographical Positions			
1	Lat	*Latitude*		
2	Long	*Longitude*		
3		*International Meridian (Greenwich)*		
4	°	*Degree(s)*		130
5	′	*Minute(s) of arc*		130
6	″	*Second(s) of arc*		130
7	PA	*Position approximate (not accurately determined or does not remain fixed)*	† (PA)	† (P.A.)	417 424.1
8	PD	*Position doubtful (reported in various positions)*	† (PD)	† (P.D.)	417 424.2
9	N	*North*		131.1
10	E	*East*		131.1
11	S	*South*		131.1
12	W	*West*		131.1
13	NE	*North-east*		
14	SE	*South-east*		
15	NW	*North-west*		
16	SW	*South-west*		

	Control Points, Distance Marks			
20	△	*Triangulation point*		304.1
21	† ⊕	*Observation spot*	† + Obs Spot	† + Obsn. Spot	304.2
22	⊙ ⊙	*Fixed point*		125.3
23	† ⼢	*Benchmark*	† ⼢ BM	† ⼢ B.M.	304.3
24		*Boundary mark*		306
25.1	○ km 32	*Distance along waterway, no visible marker*		307 361.3
25.2	○ km 32	*Distance along waterway, with visible marker*		
a		*Viewpoint*	○ See View	390.2

	Symbolised Positions (Examples)			
30	⊡ # ⌖18 Wk	*Symbols in plan: position is centre of primary symbol*		
31	⚲ ⚲ ⚲	*Symbols in profile: position is at bottom of symbol*		
32	○ Mast ⊙ MAST ★	*Point symbols (accurate positions)*		
33	† ○ Mast PA	*Approximate position*		

				Units
40	km	Kilometre(s)		
41	m	Metre(s)		130
42	dm	Decimetre(s)		130
43	cm	Centimetre(s)		
44	mm	Millimetre(s)		130
45	M	International Nautical Mile(s) (1852m) or Sea Mile(s)	n mile(s) M	130
46		Cable (0.1M)		130
47	ft	Foot/feet		
48		Fathom(s)	*fm., fms.*	
49	h	Hour		130
50	# m min	Minute(s) of time		130
51	s sec #	Second(s) of time	† sec	130
52	kn	Knot(s)		130
53	t	Tonne(s), Ton(s), tonnage (weight)		328.3
54	# cd	Candela		

				Magnetic Compass
60		Variation	Var	
61		Magnetic	Mag	
62		Bearing		132
63		true		
64		decreasing	decrg	
65		increasing	incrg	
66		Annual change		
67		Deviation		
68.1	Magnetic Variation 4°30´W 2010 (8´E) #	Note of magnetic variation, in position		272.2
68.2	Magnetic Variation at 55°N 8°W 4°30´W 2010 (8´E) #	Note of magnetic variation, out of position	Magnetic Variation: 4°30´W 2010 (10´E)	

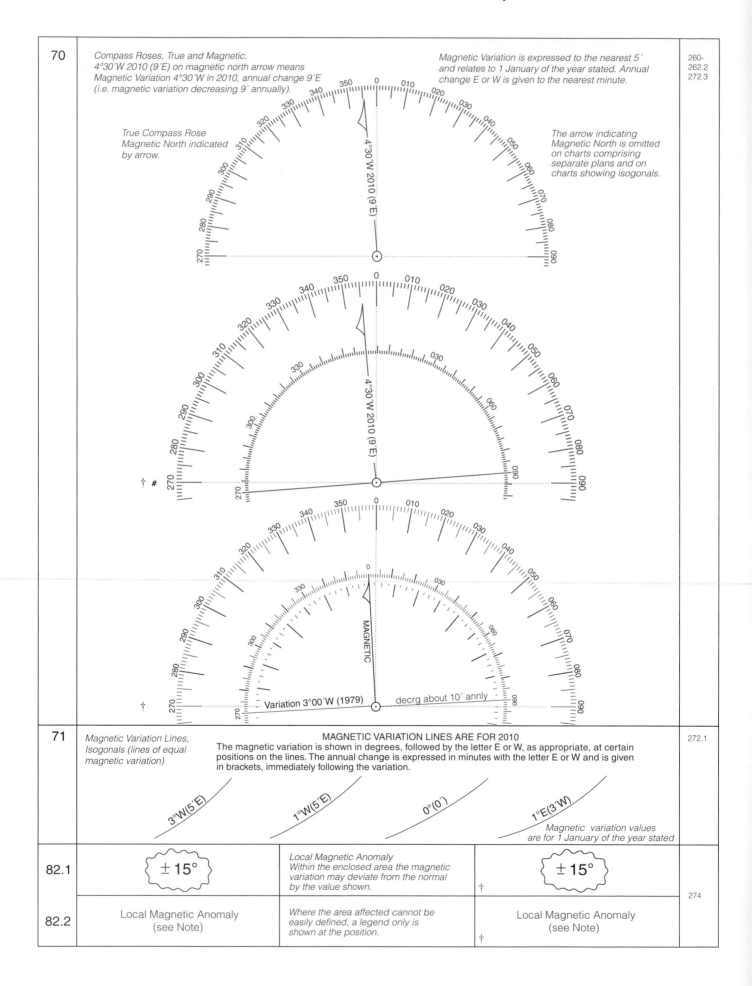

70	*Compass Roses, True and Magnetic.* *4°30´W 2010 (9´E) on magnetic north arrow means Magnetic Variation 4°30´W in 2010, annual change 9´E (i.e. magnetic variation decreasing 9´ annually).*	*Magnetic Variation is expressed to the nearest 5´ and relates to 1 January of the year stated. Annual change E or W is given to the nearest minute.*	260-262.2 272.3

True Compass Rose Magnetic North indicated by arrow.

The arrow indicating Magnetic North is omitted on charts comprising separate plans and on charts showing isogonals.

4°30´W 2010 (9´E)

4°30´W 2010 (9´E)

† #

MAGNETIC

† Variation 3°00´W (1979)   decrg about 10´ annly

71	*Magnetic Variation Lines, Isogonals (lines of equal magnetic variation)*	MAGNETIC VARIATION LINES ARE FOR 2010 The magnetic variation is shown in degrees, followed by the letter E or W, as appropriate, at certain positions on the lines. The annual change is expressed in minutes with the letter E or W and is given in brackets, immediately following the variation.	272.1

3°W(5´E)   1°W(5´E)   0°(0´)   1°E(3´W)

Magnetic variation values are for 1 January of the year stated

82.1	±15°	*Local Magnetic Anomaly Within the enclosed area the magnetic variation may deviate from the normal by the value shown.*	† ±15°	274
82.2	Local Magnetic Anomaly (see Note)	*Where the area affected cannot be easily defined, a legend only is shown at the position.*	† Local Magnetic Anomaly (see Note)	

Coastline

Foreshore → I, J

1		Coastline, surveyed	310.1 310.2
2		Coastline, unsurveyed	311
3		Steep coast, Cliffs	312.1
4		Hillocks	312.1
5		Flat coast	312.2
6		Sandy shore	312.2
7	Stones	Stony shore, Shingly shore	312.2
8	Sand dunes	Sandhills, Dunes	312.3

Relief

Plane of Reference for Heights → H

10	250 200 150 100 50 · 259 200 100	Contour lines with values and spot height	351.3 351.4 351.5 351.6 352.2
11	·437 ·359 ·189 ·115 ·49 ·123	Spot heights	352.1 352.2
12	360 300 200 100	Approximate contour lines with values and approximate height	351.3 351.4 351.5 351.6 352.3

C Natural Features

13		Form lines with spot height		351 351.2 351.3 352.2
14		Approximate height of top of trees (above height datum)		352.4

Water Features, Lava				
20		River, Stream		353.1 353.2
21		Intermittent river		353.3
22		Rapids, Waterfalls		353.5
23		Lakes		353.6
24		Salt pans		353.7
25		Glacier		353.8
26		Lava flow		355

			Vegetation			
30		W o o d e d	Woods in general			354.1
31			Prominent trees (isolated or in groups)			354.2
31.1			Deciduous tree, unknown or unspecified tree			
31.2 †			Evergreen (except conifer)			
31.3			Conifer, Casuarina			
31.4			Palm			
31.5 †			Nipa palm			
31.6 †			Casuarina			
31.7 †			Filao			
31.8 †			Eucalypt			
32			Mangrove, Nipa palm			312.4
33			Marsh, Swamp, Salt marsh, Reed beds			312.2

139

D Cultural Features

Settlements, Buildings

Height of objects → E   Landmarks → E

1			Urban area		370.3 370.4
2			Settlement with scattered buildings		370.5
3	○ Name	▢ Name #	Settlement (on medium and small-scale charts)	■ Name	370.7
4	⌖ Name	■ Name HOTEL	Inland village		370.6
5	▬ ◾ ▢ ▰ ▭		Building	Bldg	370.5
6	◆Name Hotel	◆Name Hotel	Important building in built-up area		370.3
7	NAME	NAME	Street name, Road name		371
8	Ru	⛫ Ru	Ruin, Ruined landmark	† ⛫ (ru)	378 378.2

Roads, Railways, Airfields

10			Motorway		365.1
11			Road (hard surfaced)		365.2
12	=================	----------------	Track, Path (loose or unsurfaced)		365.3
13	# #		Railway, with station	† Rly † Ry † Sta † Stn	328.4 362.1 362.2
14			Cutting	† †	363.2
15			Embankment	† †	364.1
16	⊣====⊢	⊣---⊢	Tunnel		363.1
17	✕ Airfield ⊕ Airport		Airport, Airfield		366.1 366.2
18	Ⓗ		Heliport, Helipad		366.3
a			Tramway	————————	

140

	Plane of Reference for Heights → H			Other Cultural Features		
20	20	(8.9)	Vertical clearance above Height Datum (in parentheses when displaced for clarity)	(17)† (H 17m)†	(Headway 55ft)†	380.1 380.2
21	⊢23⊣		Horizontal clearance			380.3
22	20		Fixed bridge with vertical clearance	(20)†		381.1
23.1	20		Opening bridge (in general) with vertical clearance	† (20)		381.3
23.2	7.8 Swing Bridge		Swing bridge with vertical clearance			
23.3	4.2 Lifting Bridge (open 12)		Lifting bridge with vertical clearance (closed and open)			
23.4	12 Bascule Bridge		Bascule bridge with vertical clearance			
23.5	Pontoon Bridge		Pontoon bridge	†		
23.6	5.5 Draw Bridge		Draw bridge with vertical clearance			
24	20 Transporter Bridge		Transporter bridge with vertical clearance between Height Datum and lowest part of fixed structure			381.2
25	20		Overhead transporter, Aerial cableway with vertical clearance	†	Transporter (7)	382.3
26.1	Pyl 32 Pyl		Overhead power cable with pylons and physical vertical clearance	† Power (H 30m) Power Overhead (H.98ft)		382.1
26.2	Pyl 28 Pyl		Overhead power cable with pylons and safe vertical clearance			

Note: The safe vertical clearance above Height Datum to avoid risk of electrical discharge, as defined by the responsible authority, is given in magenta where known (see H20); otherwise the physical vertical clearance is shown in black as in D20.

27	20		Overhead cable, Telephone line, with vertical clearance	† H 20m Overhead (H.64ft)		382 382.2
28	20 Overhead pipe		Overhead pipe with vertical clearance			383
29			Pipeline on land	† Pipeline		377

141

E Landmarks

General	Plane of Reference for Heights → H		Lighthouses → P	Beacons → Q	
1	◣ Factory ⊙ Hotel ▯	Examples of landmarks			340.1 340.2 340.5
2	◣ FACTORY ⊙ HOTEL ▯ WATER TOWER	Examples of conspicuous landmarks. A legend in capital letters indicates that a feature is conspicuous	†	conspic	340.1 340.2 340.3 340.5
3.1		Pictorial sketches (in true position)			373.1 390 456.5 457.3
3.2	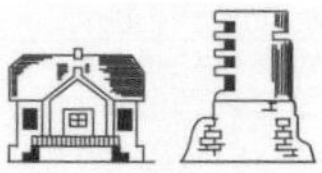	Pictorial sketches (out of position)			
4	▯ (30)	Height of top of a structure above height datum			302.3
5	▯ (30)	Height of top of a structure above ground level			303

Landmarks

Landmarks						
10.1	⊹	Ch	Church, Cathedral, prominent chapel	†	Cath	373.1 373.2
10.2	Tr	⊹ Tr	Church tower			
10.3	Sp	⊹ Sp	Church spire			373.2
10.4	Cup	⊹ Cup	Church cupola			
13	⊠		Temple, Pagoda, Shrine, Joss house	†	卍 ⊞ Pag	373.3
17	⚲		Mosque, Minaret	†	⚲	373.4
19	[L L L / L L / L L L]		Cemetery (all religions)	†	[† † † / † † / † † †] Cemy	373.6
20	▯	Tr	Tower			374.3
21	▯		Water tower, Water tank on a tower		⊙ Water Tr	374.2 376
22	▯	◣ Chy	Chimney			374.1
23	▮		Flare stack (on land)			374.1
24	▯	Mon	Monument (including column, pillar, obelisk, statue, calvary cross)	† Mont † Col ‡ #		374.4
25.1	✕		Windmill			374.5
25.2	✕ Ru		Windmill (without sails)	†	✕ (ru)	378.2

142

No.			Description				No. ref
26.1		🌀	Wind turbine	Wind turbine Windmotor	† 🌀	† 🌀	374.6
26.2	🌀	🌀	Onshore wind farm				
27	P	FS	Flagstaff, Flagpole				374.7
28		⊼	Radio mast, Television mast, Mast	⊙ Radio mast ⊙ TV mast		⊼	375.1
29		⊼	Radio tower, Television tower	⊙ Radio Tr ⊙ TV Tr			375.2
30.1	⊙ Radar Mast	⊼ Radar	Radar mast				
30.2	⊙ Radar Tr	⊼ Radar	Radar tower			⊼	487.3
30.3	⊙ Radar Sc		Radar scanner			⊼	
30.4	⊙ Radome		Radome				
31		⅄	Dish aerial	†	⊙ Dish aerial		375.4
32	▦ ⊕ •	Tanks	Tanks	†	◯		376.1 376.2
33	◯ Silo	⊙ Silo	Silo				376.3
34.1		🔺 Fort	Fortified structure (on large-scale charts)				379.1
34.2		⊞	Castle, Fort, Blockhouse (on smaller scale charts)	†	✧ Ft	Cas	379.2
34.3		⊡	Battery, Small fort (on smaller scale charts)	†	⌣ Batt	Baty	
35.1		⛏	Quarry (on large-scale charts)	†	⛏		367.1
35.2		⚔	Quarry (on smaller scale charts)				367.2
36		⚔	Mine				367.2
37.1	#	🚐	Caravan site				368
37.2		⛺	Camping site, camping and caravan site				

F Ports

Protection Structures

1		Dyke, Levee, Berm		313.1
2.1		Seawall (on large-scale charts)		313.2
2.2		Seawall (on smaller scale charts)		
3		Causeway		313.3
4.1		Breakwater (in general)		322.1
4.2		Breakwater (loose boulders, tetrapods, etc)	*(covers)*	
4.3		Breakwater (slope of concrete or masonry)		
5		Training wall		322.2
6		Groyne (always dry) Groyne (intertidal) Groyne (always underwater)		313.4 324

Harbour Installations   *Depths* → I   *Anchorages, Limits* → N   *Beacons and other fixed marks* → Q   *Marina* → U

10	⬭	Fishing harbour		320.1
11.1	⬥	Boat Harbour, Marina		
11.2	⛵	Yacht berth without facilities		320.2
11.3	▸	Yacht club, Sailing club		

12		Mole (with berthing facility)		321.3
13		Quay, Wharf	Whf	321.1
14	Pier	Pier, Jetty		321.2 321.4
15	Promenade Pier	Promenade pier		321.2
16	Pontoon	Pontoon		324.3
17	Lndg Lndg	Landing for boats	† Ldg	324.2
18		Steps, Landing stairs		324.4
19.1	④ Ⓑ (234)	Designation of berth	† ④	321.7
19.2	Ⓥ	Visitors' berth		321.8
19.3	Ⓢ	Dangerous Cargo berth		
20	◯ ▫ ▫ Dn ᖰ Dns	Dolphin		327.1
21	⚓	Deviation dolphin		327.2
22	•	Minor post or pile		327.3
23	Slip Patent slip Ramp	Slipway, Patent slip, Ramp		324.1 324.2
24		Gridiron, Scrubbing grid, Careening grid		326.8
25		Dry dock, Graving dock	†	326.1
26	Floating Dock	Floating dock	† † †	326.2
27	7·6m	Non-tidal basin, Wet dock		326.3
28		Tidal basin, Tidal harbour		326.4

F Ports

29.1	Floating Barrier - - - - - - - - - - -	Floating barrier		449.2
29.2		Oil retention barrier (high pressure pipe)		
30	Dock under construction (2011)	Works on land, with year date		329.1
31	Being reclaimed (2011)	Works at sea, Area under reclamation, with year date		329.2
32	Under construction (2011) Works in progress (2011)	Works under construction, with year date	const †constrn. †constn	329 329.4
33.1	Ru	Ruin		378.1
33.2	Pier (ru)	Ruined pier, partly submerged at high water		
34	Hulk Hulk	Hulk		
a		Bollard	∘ Bol	

Rivers, Canals, Barrages   *Clearances* → D   *Signal Stations* → T   *Cultural Features* → D

40		Canal		361
41.1	Lock	Lock (on large-scale charts)		326.6 361.6
41.2	≪	Lock (on smaller scale charts)	† ⟵	
42		Caisson, Gate		326.5
43	Flood Barrage	Flood barrage		326.7
44	Dam	Dam, Weir → Direction of flow		364.2

Transhipment Facilities   *Roads* → D   *Railways* → D   *Tanks* → E

50	RoRo	Roll-on, Roll-off (RoRo) Ferry Terminal		321.5
51	2 3 2 3	Transit shed, Warehouse (with designation)		328.1
52	♯	Timber yard		328.2
53.1	(3t)	Crane (with lifting capacity) Travelling crane on railway		328.3
53.2	(50t)	Container crane (with lifting capacity)		

Public Buildings					
60	⊕	*Harbour Master's office*	†	Hr Mr	325.1
61	⊖	*Custom office*			325.2
62.1	⊕	*Health office, Quarantine building*			325.3
62.2	⊕ Hospital	*Hospital*	⊕ Hosp † Hospl		
63	† ⊠	*Post office*	†	PO	372.1

H Tides, Currents

Terms Relating to Tidal Levels				
1	CD	Chart Datum Datum for sounding reduction		405
2	LAT	Lowest Astronomical Tide		405.3
3	HAT	Highest Astronomical Tide		
4	MLW	Mean Low Water		
5	MHW	Mean High Water		
6	MSL	Mean Sea Level		
7		Land survey datum		
8	MLWS	Mean Low Water Springs		
9	MHWS	Mean High Water Springs		
10	MLWN	Mean Low Water Neaps		
11	MHWN	Mean High Water Neaps		
12	MLLW	Mean Lower Low Water		
13	MHHW	Mean Higher High Water		
14	MHLW	Mean Higher Low Water		
15	MLHW	Mean Lower High Water		
16	Sp	Spring tide	† Spr.	
17	Np	Neap tide		
a		High Water	HW	
b		Low Water	LW	
c		Mean Tide Level	MTL	
d		Ordnance Datum	OD	

Vertical clearance → D	Tide Gauge → T	Tidal Levels and Charted Data

20 NOTE: Planes of reference are not exactly as shown below for all charts. They are usually defined in notes under chart titles.

302.2
380.1
405

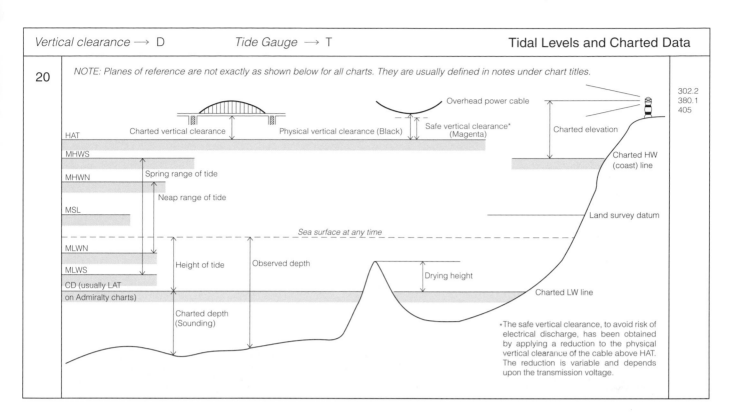

*The safe vertical clearance, to avoid risk of electrical discharge, has been obtained by applying a reduction to the physical vertical clearance of the cable above HAT. The reduction is variable and depends upon the transmission voltage.

Tide Tables

30 *Tabular statement of semi-diurnal or diurnal tides*

406.2
406.3
406.4
406.5

Tidal Levels referred to Datum of Soundings

Place	Lat. N/S	Long. E/W	Heights in metres/feet above datum				Datum and Remarks
			MHWS	MHWN	MLWN	MLWS	
			MHHW	MLHW	MHLW	MLLW	

31 *Tidal stream table*

407.2
407.3

Tidal streams referred to....

Hours	◇ Geographical Position				Ⓐ	Ⓑ	Ⓒ	Ⓓ	Ⓔ
	Directions of streams (degrees)	Rates at spring tides (knots)	Rates at neap tides (knots)						

Before High Water	6
	5
	4
	3
	2
	1
High Water	
After High Water	1
	2
	3
	4
	5
	6

	-6
	-5
	-4
	-3
	-2
	-1
	0
	+1
	+2
	+3
	+4
	+5
	+6

Ⓔ
No
Maximum Rates
For predictions, use Admiralty Tide Tables

H Tides, Currents

			Tidal Streams and Currents	Breakers →K	Tide Gauge → T	
40	*3kn →*		Flood tide stream (with mean spring rate)	†	The number of black dots on the tidal stream arrows indicates the number of hours after High or Low Water at which the streams are running	407.4 408.2
41	*2, 8kn →*		Ebb tide stream (with mean spring rate)	†		407.4 408.2
42	# ≫≫→ ~~~→		Current in restricted waters	† ≫≫→		408.2
43	~~~→ (see Note)		Ocean current. Details of current strength and seasonal variations may be shown			408.3
44			Overfalls, tide rips, races	†		423.1
45			Eddies			423.3
46	◈Ⓓ		Position of tabulated tidal stream data with designation	† ◈Ⓓ		407.2
47	[a]		Offshore position for which tidal levels are tabulated			406.5
e			Wave recorder (see L25)	† ⊙ Wave recorder		
f			Current meter (see L25)	† ⊙ Current meter		

		General		
1	*ED*	Existence doubtful	† *(ED)*	417 424.3
2	*SD*	Sounding of doubtful depth		417 424.4
3.1	*Rep*	Reported, but not confirmed	† *Repd*	417 424.5
3.2	*Rep (1973)*	Reported, with year of report, but not confirmed	† *Repd (1973)*	
4	:184: :212:	Reported, but not confirmed, sounding or danger (on small-scale charts only)		S-4 Part C 404.3
a		Unexamined	*unexam* †*unexamd*	

	Plane of Reference for Depths → H	*Plane of Reference for Heights → H*	Soundings and Drying Heights	
10	12 9_2 # $9,7$	Sounding in true position		403.1 410/412 412.1
11	• (4_8) + (12) ⊚ 3349	Alongside depth, Sounding out of position	• (8_3) $\overline{(10_4)}$ + 1_8 8_7 $\underline{7}_1$ #	412 412.1 412.2
12	$)($ (14_7)	Least depth in narrow channel		412 412.1 412.2
13	$\overset{\cdot}{\underline{330}}$	No bottom found at depth shown		412.3
14	12 9_1	Soundings which are unreliable (e.g. taken from old or smaller scale sources) shown in upright, hairline figures		412.4 417.3
15	4_9 4 0_9 3_4 2 0	Drying heights and contours above chart datum		413 413.1 413.2
16	1_4 0 2_5 0_6 1_7 2_7	Natural watercourse (in intertidal area)		413.3

	Plane of Reference for Depths → H		Depths in Channels and Areas	
20	------- -------	Limit of dredged channel or area (major and minor)	# _____	414.3
21	7·0m 3·5m	Dredged channel or area with minimum depth regularly maintained	Depths may be shown as *3,5* or 3_5 on some adopted charts. There may be a note to clarify the maintenance regime	414
22	17m (2011) 8·2m (2011)	Dredged channel or area with minimum depth not regularly maintained and year of latest survey		414.1

I Depths

24		Area swept by wire drag. The depth is shown at Chart Datum. (The latest date of sweeping may be shown in parentheses)		415 415.1
			†	

25		Unsurveyed or inadequately surveyed area; area with inadequate depth information		410 417 417.6 417.7 418

Depth Contours

30		Dries 2m contour Low Water (LW) Line, Chart Datum (CD) Blue tint, in one or more shades, and tint ribbons, are shown to different limits according to the scale and purpose of the chart and the nature of the bathymetry. On some charts, the standard set of contours is augmented by additional contours in order to delimit particular bathymetric features or for the benefit of particular categories of shipping. However, in some instances where the provision of additional contours would be helpful, the survey data available does not permit it. On charts which are metric conversions of fathoms charts there will be a non-standard series of contours. On some charts, contours are printed in blue.	On charts showing depths in fathoms/feet, the following contours are used: On some recently-corrected charts, contours may be shown by continuous lines.	404.2 410 411

31		Approximate depth contours (length of dashes may vary)		411.2 417.5

	Rocks →K		Types of Seabed		
1	S	Sand	†	s	425-427
2	M	Mud	†	m	
3	Cy	Clay	†	cl	
4	Si	Silt			
5	St	Stones	†	st	
6	G	Gravel	†	g	
7	P	Pebbles	†	peb	
8	Cb	Cobbles			
9.1	R	Rock, Rocky	†	r	
9.2	Bo	Boulder(s)			421.2 425-427
10	Co	Coral	†	crl	425-427
11	Sh	Shells	†	sh	
12.1	S/M	Two layers e.g. Sand over Mud	#M (25)/SG S (<1)/R (Thickness of surface layer in metres)		425.8
12.2	fS.M.Sh	Mixed: where the seabed comprises a mixture of materials, the main constituent is given first, e.g. fine Sand with Mud and Shells			425.9
13.1	Wd	Weed (including Kelp)	†	wd	425.5
13.2	⟨kelp symbol⟩	Kelp			428.2
14	⟨sandwaves symbol⟩	Sandwaves			428.1
15	⟨spring symbol⟩	Spring in seabed			428.3
a		Ground	†	Gd grd	
b		Ooze	†	Oz	
c		Marl	†	Ml	
d		Shingle	†	Sn shin	
e		Chalk	†	Ck chk	
f		Quartz	†	Qz qrtz	
g		Madrepore	†	Md mad	
h		Basalt	†	Ba	
i		Lava	†	Lv	
j		Pumice	†	Pm pum	
k		Tufa	†	T	
l		Scoriæ	†	Sc	
m		Cinders	†	Cn cin	

n		Manganese	†	Mn	man
o		Glauconite	†	Gc	
p		Oysters	†	Oy	oys
q		Mussels	†	Ms	mus
r		Sponge	†	Sp	
s		Algae	†	Al	
t		Foraminifera	†	Fr	for
u		Globigerina	†	Gl	
v		Diatoms	†	Di	
w		Radiolaria	†	Rd	rad
x		Pteropods	†	Pt	
y		Polyzoa	†	Po	pol

Intertidal Areas

20		Area of sand and mud with patches of stones or gravel			426.1
21		Rocky area			426.2
22		Coral reef			426.3

Qualifying Terms

30	f	Fine			425 427
31	m	Medium — only used in relation to sand			
32	c	Coarse			
33	bk	Broken	†	brk	
34	sy	Sticky	†	stk	
35	so	Soft	†	sft	
36	sf	Stiff	†	stf	
37	v	Volcanic	†	vol	
38	ca	Calcareous	†	cal	
39	h	Hard			425.5 425.7

aa		Small	†		*sm*		
ab		Large	†		*l*		
ac		Glacial	†	*ga*		glac	
ad		Speckled	†	sk		spk	
ae		White	†		w		
af		Black	†	bl		blk	
ag		Blue	†		b		
ah		Green	†		gn		
ai		Yellow	†		y		
aj		Red	†		rd		
ak		Brown	†		br		
al		Chocolate	†	ch		choc	
am		Grey	†		gy		
an		Light	†		lt		
ao		Dark	†		d		

K Rocks, Wrecks, Obstructions, Aquaculture

General				
1		*Dangerline: A danger line draws attention to a danger which would not stand out clearly enough if represented solely by its symbol (e.g. isolated rock) or delimits an area containing numerous dangers, through which it is unsafe to navigate*		411.4 420.1
2	7_5	*Depth cleared by wire drag sweep or examined by diver. The symbol may be used with other symbols, e.g. wrecks, obstructions, wells*		415 422.3 422.9
3	(12)	*Safe clearance depth. Obstruction over which the exact depth is unknown, but which is estimated to have a safe clearance at the depth shown. The symbol may be used with other symbols, e.g. wrecks, wells, turbines*		422.5 422.7 422.9
a		*Dries*	† Dr † dr	
b		*Covers*	† cov	
c		*Uncovers*	† uncov	

Rocks	*Plane of Reference for Heights* → H		*Plane of Reference for Depths* → H	
10		*Rock (islet) which does not cover, height above height datum*	(1,7) (3,1) (4,1)	421.1
11		*Rock which covers and uncovers, height above Chart Datum, where known*		421.2
12		*Rock awash at the level of Chart Datum*		421.3
13		*Underwater rock over which the depth is unknown, but which is considered dangerous to surface navigation*		421.4
14		*Underwater rock of known depth:*		421.4
14.1		*inside the corresponding depth area*		
14.2		*outside the corresponding depth area, dangerous to surface navigation*		

15	35 R	Underwater rock of known depth, not dangerous to surface navigation			421.4
16	+ +Co + + Co 5₈	Coral reef which is always covered			421.5
17	5₈ Br 18 19	Breakers			423.2
d		Discoloured water	Discol	† Discold	424.6

	Hulk → F	Plane of Reference for Depths → H	Historic Wreck → N		**Wrecks and Fouls**
20	Mast (1·2) Wk	Wreck, hull never covers, on large-scale charts			422.1
21	Mast (1₂) Wk	Wreck, hull covers and uncovers, on large-scale charts	Wk †	Wk †	
22	5₂ Wk 6₅ Wk	Submerged wreck, depth known, on large-scale charts	5₂ Wk †		422.1
23	Wk	Submerged wreck, depth unknown, on large-scale charts	Wk †		422.1
24		Wreck showing any part of hull or superstructure at the level of Chart Datum			422.2
25	Masts	Wreck of which the mast(s) only are visible at Chart Datum	Mast (1·2) Wk Funnel Mast (1₂)		422.2
26	4₆ Wk 25 Wk	Wreck over which the depth has been obtained by sounding but not by wire sweep			422.4
27	4₆ Wk 25 Wk	Wreck, least depth obtained by wire sweep or diver			422.3
28		Wreck, depth unknown, which is considered potentially dangerous to surface navigation			422.6
29	+++	Wreck, in over 200m or depth unknown, which is considered not dangerous to surface navigation. For information about depth criteria, which may vary, see NP100, The Mariner's Handbook			422.6
e		Submerged wreck, depth unknown	Wk †		

30	20 Wk	Wreck over which the exact depth is unknown, but which is estimated to have a safe clearance at the depth shown		422.5 422.7
31.1	# # (22)	Foul ground, not dangerous to surface navigation, but to be avoided by vessels anchoring, trawling, etc. (e.g. remains of wreck, cleared platform). Foul ground with depth	† ◯ Foul † 22 Foul	422.8
31.2	# # (#)	Area of foul ground	† ⬚ Foul † Foul	
f		Navigation light on stranded wreck		470.5

Obstructions and Aquaculture		Plane of Reference for Depths → H Underwater Installations → L	Kelp, Seaweed → J	
40	Obstn Obstn	Obstruction or danger to navigation the exact nature of which is not specified or has not been determined, depth unknown		422.9
41	4₆ Obstn 16₈ Obstn	Obstruction, depth obtained by sounding but not wire sweep		422.9
42	4₆ Obstn 16₈ Obstn	Obstruction, least depth obtained by wire sweep or diver		422.9
43.1	Obstn ⊤ ⊤ ⊤ #	Stumps of posts or piles, wholly submerged		422.9
43.2	⊤ #	Submerged pile, stake, snag or stump (with exact position)		
44.1		Fishing stakes	† ⊥ ⊥ ⊥ ⊥ ⊥ †	447.1
44.2		Fish trap, fish weir, tunny nets	†	447.2
45	Fish traps Tunny nets	Fish trap area, tunny nets area		447.3
46.1		Fish haven		447.5
46.2	2₄ (2₄)	Fish haven, with minimum depth		
47		Shellfish beds	Shellfish Beds †	447.4
48.1		Marine farm (on large-scale charts)	† Fish farm † Fish cages	447.6
48.2		Marine farm (on small-scale charts)		

	Combined symbols → K (General)	Areas, Limits → N		General
1	*EKOFISK OILFIELD*	Name of oilfield or gasfield		445.3
2	⊡ Z-44	Platform with designation/name	† ★ † ⊡	445.3
3		Limit of safety zone around offshore installation		439.2 445.6
4		Limit of development area		445.7
5.1		Wind turbine, floating wind turbine and wind turbine with vertical clearance		445.8
5.2		Offshore wind farm		445.9
		Offshore wind farm (floating)		
6		Wave farm, Renewable energy device		445.12

	Mooring Buoys → Q			Platforms and Moorings
10		Production platform, Platform, Oil derrick	† ★ † ⊡	445.2
11	⊡ Fla	Flare stack (at sea)		445.2
12	⊡ SPM	Fixed Single Point Mooring, including Single Anchor Leg Mooring (SALM), Articulated Loading Column (ALC)		445.2 445.4
13		Observation / research platform (with name)	⊡ Name	
14	⊡ Ru ⊡ Z-44 (ru)	Disused platform, with superstructure removed		445.2
15		Artificial Island	Name	
16		Floating Single Point Mooring, including Catenary Anchor Leg Mooring (CALM), Single Buoy Mooring (SBM)		445.4
17		Moored storage tanker including FSO, FSU and FPSO, Accommodation vessel		445.5
18		Mooring ground tackle for fixing floating structures		431.6

	Plane of Reference for Depths → H	Obstructions → K		Underwater Installations
20	15 Well Well	Production well, with depth where known	† Prod Well	445.1
21.1	Well	Suspended well (wellhead and pipes projecting from the seabed) over which the depth is unknown		445.1
21.2	15 Well	Suspended well over which the depth is known		445.1
21.3		Suspended well with height of wellhead above the sea floor	# Well (5.7)	

L Offshore Installations

22	#		Site of cleared platform		422.8
23	✦ ⊙ Pipe	⊙ Pipe (1₈)	Above-water wellhead (lit and unlit). The drying height or height above height datum is charted if known		445.1
24	⊙ Turbine	Fl(2) ☆ Underwater Turbine	Underwater turbine		445.10 445.11
25	⊙ ODAS		Subsurface Ocean (or oceanographic) Data Acquisition System (ODAS)		448.4

Submarine Cables

30.1	〜〜〜〜〜〜〜〜〜	Submarine cable	† 〜〜〜〜〜〜〜〜〜	443.1 443.8
30.2	⊤⊤⊤⊤〜〜〜〜〜⊤⊤⊤⊤ ⊥⊥⊥⊥〜〜〜〜〜⊥⊥⊥⊥	Submarine cable area	† – – – – Cable Area – – – –	439.3 443.3
31.1	〜〜〜〜ʃ〜〜〜〜	Submarine power cable	† 〜〜〜〜Power〜〜〜〜 † 〜〜〜〜Power〜〜〜〜	443.2
31.2	⊤⊤⊤⊤〜〜ʃ〜〜⊤⊤⊤⊤ ⊥⊥⊥⊥〜〜ʃ〜〜⊥⊥⊥⊥	Submarine power cable area	† – – – Power Cable Area – – –	439.3 443.2
32	〜〜〜〜〜〜〜〜	Disused submarine cable		443.7

Submarine Pipelines

40.1	→→→→→→→→→→→ Oil Gas →→→→→→ →→→→→→ Chem Water	Supply pipeline: unspecified, oil, gas, chemicals, water	† – – – – Pipeline – – – –	444 444.1
40.2	→→→⊤⊤⊤⊤⊤⊤⊤⊤→→→ ←←←⊥⊥⊥⊥⊥⊥⊥⊥←←← →→→⊤⊤⊤⊤ ⊤⊤⊤→→→ Oil Gas ←←←⊥⊥⊥⊥ ⊥⊥⊥←←← →→→⊤⊤⊤⊤ ⊤⊤⊤→→→ Chem Water ←←←⊥⊥⊥⊥ ⊥⊥⊥←←←	Supply pipeline area: unspecified, oil, gas, chemicals, water	† ⌐ Pipeline Area ⌐ † ⌐ Pipeline Area ⌐	439.3 444.3
41.1	→→→→→→→→→→ Water Sewer →→→→→ →→→→→ Outfall Intake →→→→→ →→→→→	Outfall and intake: unspecified, water, sewer, outfall, intake	† – – – – Sewer – – – – † – – – – Outfall – – – –	444 444.2
41.2	→→→⊤⊤⊤⊤⊤⊤⊤⊤⊤→→→ ←←←⊥⊥⊥⊥⊥⊥⊥⊥⊥←←← →→→⊤⊤⊤ ⊤⊤⊤→→→ Water Sewer ←←←⊥⊥⊥ ⊥⊥⊥←←← →→→⊤⊤⊤ ⊤⊤⊤→→→ Outfall Intake ←←←⊥⊥⊥ ⊥⊥⊥←←←	Outfall and intake area: unspecified, water, sewer, outfall, intake	† ⌐ Pipeline Area ⌐ † ⌐ Pipeline Area ⌐	439.3 444.3
42.1	→ → → Buried 1·6m → → →	Buried pipeline / pipe (with nominal depth to which buried)		
42.2	→→→) (←→→	Pipeline tunnel		444.5
43	→→→→→→→→→ ⊙ 3₂ Obstn	Diffuser, crib (nature of obstruction be stated)		444.8
44	→→→ →→→ →→→ → → → → →→→ →→→ →→→ → → →	Disused pipeline / pipe		444.7

160

Tracks Marked by Lights → P		*Leading Beacons → Q*		Tracks

1	270·5° 2 Bns ≠ 270·5°	Leading line (≠ means "in line", the continuous line is the track to be followed)	† Bn　Bn　*Bns in Line 270°30′* Ldg Bns 270·5° 270·5°	433.1 433.2 433.3
2	270·5° Island open of Headland 270·5°	Transit (other than leading line), Clearing line	Bns in line 270·5°	433.4 433.5
3	090°-270°	Recommended track based on a system of fixed marks ‡	† † †	434.1 434.2
4	090°-270° <—>	Recommended track not based on a system of fixed marks ‡	— — ‹ — — DW — 270° — — — — ‹ — — —	434.1 434.2
5.1	DW (see Note)	One-way track and DW track based on a system of fixed marks	† †	432.3 434.1
5.2	270° DW	One-way track and DW track not based on a system of fixed marks		
6	‹ 7·3m › ‹ 7·3m ›	Recommended track with maximum authorised draught ‡		432.4 434.3 434.4

				Routeing Measures - Basic Symbols

10	⇒	Established (mandatory) direction of traffic flow		435.1
11	⇢	Recommended direction of traffic flow ‡		435.5
12		Separation line (large-scale, small-scale)		435.1 436.3
13		Separation zone		435.1 436.3
14		Limit of restricted routeing measure (e.g. Inshore Traffic Zone, Area to be Avoided)		435.1 436.3 439.2
15		Limit of routeing measure		435.1 436.3
16	⚠	Precautionary Area	Precautionary area		435.2
17	ASL (see Note)	Archipelagic Sea Lane; axis line and limit beyond which vessels shall not navigate	ASL (see Note) †	435.10
18	FAIRWAY 7·3m FAIRWAY <7·3m>	Fairway, designated by regulatory authority: with minimum depth with maximum authorised draught	FAIRWAY 7·3m FAIRWAY <7·3m> #	434.5

‡ *The term 'recommended' in connection with tracks and routeing measures does not imply recommendation by the United Kingdom Hydrographic Office. It is usually by a regulatory authority, but may be established by precedent.*

M Tracks, Routes

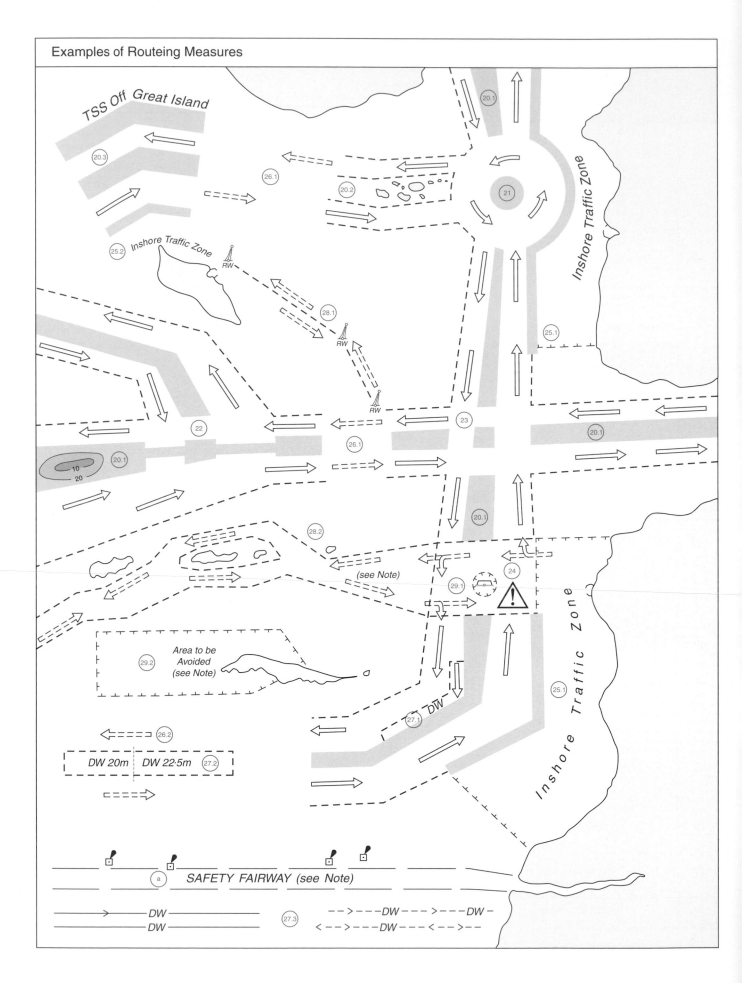

Examples of Routeing Measures (see diagram on page 34)

(20.1)	Traffic separation scheme (TSS), traffic separated by separation zone	435.1
(20.2)	Traffic separation scheme, traffic separated by natural obstructions	435.1
(20.3)	Traffic separation scheme, with outer separation zone, separating traffic using scheme from traffic not using it	435.1
(21)	Traffic separation scheme, roundabout	435.1
(22)	Traffic separation scheme with "crossing gates"	435.1
(23)	Traffic separation schemes crossing, without designated precautionary area	435.1
(24)	Precautionary area	435.2
(25.1)	Inshore traffic zone (ITZ), with defined end limits	435.1
(25.2)	Inshore traffic zone, without defined end limits	435.1
‡ (26.1)	Recommended direction of traffic flow, between traffic separation schemes	435.5
‡ (26.2)	Recommended direction of traffic flow, for ships not needing a deep water route	435.5
(27.1)	Deep water route (DW), as part of one-way traffic lane	435.3
(27.2)	Two-way deep water route, with minimum depth stated	435.3
(27.3)	Deep water route, centre line shown as recommended one-way or two-way track	435.3
‡ (28.1)	Recommended route (often marked by centre line buoys)	435.4
(28.2)	Two-way route with one-way sections	435.6
(29.1)	Area to be avoided (ATBA), around aid to navigation	435.7
(29.2)	Area to be avoided, because of danger of stranding	435.7
(a)	Safety fairway (U.S. waters)	

‡ The term 'recommended' in connection with tracks and routeing measures does not imply recommendation by the United Kingdom Hydrographic Office. It is usually by a regulatory authority, but may be established by precedent.

Radar Surveillance System

30	⊙ Radar Surveillance Station	Radar surveillance station		487 487.3
31	Ra Cuxhaven	Radar range		487.1
32.1	– – – – – – – – Ra – – – – – – – –	Radar reference line		487.2
32.2	Ra 090° - 270°	Radar reference line coinciding with a leading line		

Radio Reporting

40.1	[B] [7] VHF 80	Radio calling-in point, way point, or reporting point (with designation, if any) showing direction(s) of vessel movement and VHF-channel	† [B] [7]	488.1
40.2	– ◊ – – – – – – – – – – – – ◊ – –	Radio reporting line (with designation, if any) showing direction(s) of vessel movement	† ◊ † ◊	488.2

Ferries

50	– – – – – – – – – ⊙ – – – – – – – – –	Ferry Route	† – – – – – Ferry – – – – – – † – – – – – Ferry – – – – – –	438.1
51	– – – – – Cable Ferry – – – – – ⊙ – – – – –	Cable Ferry Route		438.2

N Areas, Limits

General	Dredged and Swept Areas → I	Submarine Cables, Submarine Pipelines → L	Tracks Routes → M	
1.1	(for emphasis)	Maritime limit in general, usually implying permanent physical obstructions		439.1 439.6
1.2	(for emphasis)	Maritime limit in general, usually implying no permanent physical obstructions		
2.1	(for emphasis)	Limit of restricted area		439.2- 439.4 439.6 441.6
2.2		Limit of area into which entry is prohibited	† Entry Prohibited	

Note: On multicoloured charts, these symbols may be in green when associated with environmental areas

Anchorages, Anchorage Areas				
10	⚓	Reported anchorage (no defined limits)	† ⚓	431.1
11.1	(A) (N53) (14)	Anchor berths	† N53	431.2
11.2	(A) (N53) (14)	Anchor berths with swinging circle	† (N53)	
12.1	⚓	Anchorage area in general. On smaller scale charts, the limits may be omitted		431.3
12.2	No 1 ⚓	Numbered anchorage area	† (1) † (1)	
12.3	Oaze ⚓	Named anchorage area		
12.4	DW ⚓	Deep water anchorage area, anchorage area for deep-draught vessels		
12.5	Tanker ⚓	Tanker anchorage area. This symbol may be adapted for other types of vessel, e.g. small craft		
12.6	24h ⚓	Anchorage area for periods up to 24 hours		
12.7	⚓	Dangerous cargo anchorage area		
12.8	⊕ ⚓	Quarantine anchorage area		
12.9	Reserved (see Note) ⚓	Reserved anchorage area		
13	✈	Seaplane operating area	†	449.6
14	⚓	Anchorage for seaplanes	† ⚓	449.6

Restricted Areas

20		Anchoring prohibited	† [Anchoring Prohibited] † [☠]	431.4 435.11 439.3 439.4
21.1		Fishing prohibited		
21.2		Diving prohibited		439.3 439.4
22	Examples: ⊤⊤⊤ MR ⊤⊤⊤ MR ⊤⊤⊤ ⊤⊤⊤ MR ⊤⊤⊤⊤ MR ⊤⊤⊤	Environmentally Sensitive Sea Areas: (colour may be green or magenta) Limit of marine reserve, national park, non-specific nature reserve	† [Marine Nature Reserve (see Note)]	437.3 437.6 437.7
	Examples (bird/seal silhouettes)	Bird sanctuary, Seal sanctuary (other animal silhouettes may be used for specialized areas)	# — — — 𝒮 — — — — — 𝒮 — —	
	PSSA PSSA	Particularly Sensitive Sea Area (coloured tint band may vary in width between 1 and 5mm)		
23.1	Explosives Dumping Ground	Explosives dumping ground, Individual mine or explosive	† [Explosives Dumping Ground]	442.1 442.2 442.3 442.4
23.2	Explosives Dumping Ground (disused)	Explosives dumping ground (disused)	† [Explosives Dumping Ground (disused)]	
24	Dumping Ground for Chemicals	Dumping ground for chemical waste		442.1 442.2 442.3
25	Degaussing Range	Degaussing range	† D.G. Range DG Range	448.1 448.2
27	5kn	Maximum speed, Speed limit		430.2
a		Seabed operations dangerous/prohibited	# [⚓ ⚔ ⚫ ☠ ⚔ ⚫]	

Military Practice Areas

30		Firing practice area		441.1 441.2 441.3
31	❶ [⬤]	Military restricted area into which entry is prohibited	† [Entry Prohibited]	441.6
32	○⊦ [○ ○]	Mine-laying (and counter-measure) practice area		441.4
33	SUBMARINE EXERCISE AREA	Submarine transit lane and exercise area		441.5
34	Minefield	Minefield	[Mine Danger Area (see Note)]	441.8

N Areas, Limits

International Boundaries and National Limits

40	DANMARK + DEUTSCHLAND	International boundary on land	† DENMARK + + + + + + + + + + + + + + + + + + + GERMANY — · — · — · — · — · — · — · — · — · — · —	440.1
41	UNITED KINGDOM — + — + — + — + — + — + — + — + NORGE	International maritime boundary	† UNITED KINGDOM — + — + — + — + — + — + — + NORWAY † Continental Shelf — — — — — — — — Boundary — — — —	440.3
42		Straight territorial sea baseline with base point		440.4
43	——————— + + ———————	Seaward limit of Territorial Sea	# +	440.5
44	——————— + ———————	Seaward limit of Contiguous Zone		440.6
45	— ×⊏⊳ — — — — ×⊏⊳ — — ×⊏⊳ — — — — — ×⊏⊳ —	National fishery limits		440.7
46	——— Continental Shelf ———	Limit of Continental Shelf		440.8
47	——— EEZ ———	Limit of Exclusive Economic Zone	# ——— + + ——————— + + ———	440.9
48	— — ⊖ — — — — — — ⊖ — —	Customs limit		440.2
49	Harbour Limit	Harbour limit	† Harbour Limit	430.1

Various Limits

60.1	(2008) #	Limit of fast ice, Ice front (with date)	†	449.1
60.2	(2008) #	Limit of sea ice (pack ice) seasonal (with date)		
61	Log Pond	Floating barrier, including log ponds, security barriers, ice booms, shark nets	† Booming Ground † Timber	449.2
62.1	Spoil Ground	Spoil ground		446.1 446.2
62.2	Spoil Ground (disused)	Spoil ground (disused)		
63	Extraction Area	Extraction (dredging) area	† Dredging Area	446.4
64	Cargo Transhipment Area	Cargo transhipment area		449.4
65	† Incineration Area	Incineration area	† Area for burning refuse material	449.3

	Beacons → Q				Light Structures, Major Floating Lights		
1.1	☆	★	Lt	LtHo	Position of navigation light (size and style of 'star' may vary), light, lighthouse	# ❗	
1.2		✦			Light on standard charts		470.4 470.5
1.3		⦾			Significant all-round light on multicoloured charts (generally for offshore navigation)		

Note: On standard charts, positions of light are highlighted by one magenta flare.
On multicoloured charts, the flare indicates the colour of the light, except for multicoloured sector lights where a magenta flare may be used if the sectors are not charted.
This guide shows standard magenta flares with examples of multicoloured depiction where significantly different.

2			▣✦		Lighted offshore platform		445.2
3	BY		✦ BnTr		Lighted beacon tower ‡	† 🔲Bn Tower † Bn Tr	456.4 457.1 457.2
4	✦	R	BRB	✦ Bn	Lighted beacon ‡ On smaller scale charts, where navigation within recognition range of the daymark is unlikely, lighted beacons are charted solely as lights	# R BRB ☆ G ☆ R	457.1 457.2
5	R		✦ Bn		Lighted buoyant beacon, resilient beacon ‡		459.1 459.2
6			⬯✦		Large Automatic Navigational Buoy (LANBY))	† ⬯	462.9 474
7	⬚✦	⬭✦	ⵍ✦	▣✦	Navigation lights on landmarks or other structures (examples)		470.5

‡ Minor lights, fixed and floating, usually conform to IALA Maritime Buoyage System characteristics

Bearings of Light Off Chart Limits

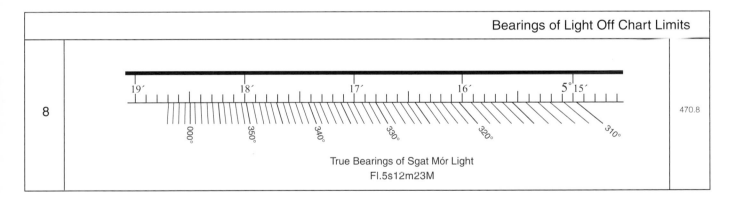

True Bearings of Sgat Mór Light
Fl.5s12m23M

8		470.8

P Lights

	Abbreviation		Class of Light	Illustration Period shown ⊢——⊣	
	International	National			
10.1	F		Fixed		
10.2	Occulting (total duration of light longer than total duration of darkness)				
	Oc	Occ †	Single-occulting		
	Oc(2) *Example*	GpOcc(2) *Example* †	Group-occulting		
	Oc(2+3) *Example*	GpOcc(2+3) *Example* †	Composite group-occulting		
10.3	Isophase (duration of light and darkness equal)				
	Iso		Isophase		
10.4	Flashing (total duration of light shorter than total duration of darkness)				
	Fl		Single-flashing		
	Fl(3) *Example*	GpFl(3) *Example* †	Group-flashing		
	Fl(2+1) *Example*	GpFl(2+1) *Example* †	Composite group-flashing		
10.5	LFl		Long-flashing (flash 2s or longer)		
10.6	Quick (repetition rate of 50 to 79 - usually either 50 or 60 - flashes per minute)				
	Q	QkFl †	Continuous quick		
	Q(3) *Example*	QkFl(3) *Example* †	Group quick		
	IQ †	IntQkFl †	Interrupted quick		
10.7	Very quick (repetition rate of 80 to 159 - usually either 100 or 120 - flashes per minute)				
	VQ	VQkFl †	Continuous very quick		
	VQ(3) *Example*	VQkFl(3) *Example* †	Group very quick		
	IVQ †	IntVQkFl †	Interrupted very quick		
10.8	Ultra quick (repetition rate of 160 or more - usually 240 to 300 - flashes per minute)				
	UQ		Continuous ultra quick		
	IUQ		Interrupted ultra quick		
10.9	Mo(K) *Example*		Morse Code		
10.10	FFl		Fixed and flashing		
10.11	Al.WR *Example*	Alt.WR *Example* †	Alternating		

Colours of Lights and Marks

11.1	W		White (for lights, only on sector and alternating lights)		450.2 450.3 470.4
11.2	R		Red		470.6 471.3
11.3	G		Green		471.4 475.1
11.4	Bu		Blue	† Bl	
11.5	Vi		Violet		
11.6	Y		Yellow		
11.7	Y #	Or	Orange	† Or	
11.8	Y #	Am	Amber		

		Colours of lights shown on:
		standard charts
		on multicoloured charts
		on multicoloured charts at sector lights

Period

12	90s 2·5s *Examples*	Period in seconds and tenths of a second	† 90sec	471.5

Plane of Reference for Heights → H   *Tidal Levels* → H

Elevation

13	12m *Example*	Elevation of light given in metres	On fathoms charts, the elevation of a light is given in feet e.g. **40ft**	471.6

Range

Note: Charted ranges are nominal ranges given in sea miles

14	15M *Example*	Light with single range		471.7 471.9 475.5
	15/10M *Example*	Light with two different ranges	† 15,10M	
	15-7M *Example*	Light with three or more ranges	† 15,10,7M	

Disposition

15	(hor)	horizontally disposed	† (horl.)	
	(vert)	vertically disposed	† (vertl.)	471.8
	(△) (▽)	3 lights disposed in the shape of a triangle		

Example of a full Light Description   471.9

16	**Example** *of a light description on a metric chart using international abbreviations:* ★ Fl(3)WRG.15s13m7-5M	**Example** *of a light description on a fathoms chart using international abbreviations:* ★ Al.Fl.WR.30s110ft23/22M
	Fl(3) — **Class** or **character** of light: in this example a group-flashing light, regularly repeating a group of three flashes.	**Al.Fl.** — **Class** or **character** of light: in this example exhibiting single flashes of differing colours alternately.
	WRG. — **Colours** of light: white, red and green, exhibiting the different colours in defined sectors.	**WR.** — **Colours** of light shown alternately: white and red all-round (i.e. not a sector light).
	15s — **Period** of light in seconds, i.e., the time taken to exhibit one full sequence of 3 flashes and eclipses: 15 seconds.	**30s** — **Period** of light in seconds, i.e. the time taken to exhibit the sequence of two flashes and two eclipses: 30 seconds.
	13m — **Elevation** of focal plane above height datum: 13 metres.	**110ft** — **Elevation** of focal plane above height datum: 110 feet.
	7-5M — **Luminous range** in sea miles: the distance at which a light of a particular intensity can be seen in 'clear' visibility, taking no account of earth curvature. In those countries (e.g. United Kingdom) where the term 'clear' is defined as a meteorological visibility of 10 sea miles, the range may be termed "nominal". In this example the ranges of the colours are: white 7 miles, green 5 miles, red between 7 and 5 miles.	**23/22M** — **Range** in sea miles. Until 1971 the lesser of **geographical** range (based on a height of eye of 15 feet) and **luminous** range was charted. Now, when the charts are corrected, luminous (or nominal) range is given. In this example the luminous ranges of the colours are: white 23 miles, red 22 miles. The geographical range can be found from the table in the ADMIRALTY List of Lights (for the elevation of 110 feet, it would be 16 miles).

P Lights

Lights marking Fairways	Note: Quoted bearings are always from seaward

Leading Lights and Lights in line

20.1	Oc.3s8m12M ☆ Oc.225.3° Oc.6s Oc.3s ★ Oc.6s24m15M	Leading lights with leading line (the firm line is the track to be followed) and arcs of visibility	Oc.3s8m12M ☆ Oc.6s Oc.3s ★ Oc.6s24m15M	433 433.1 433.2 433.3 475.1 475.6
20.2	Oc.4s12M ☆ Oc.R. 4s10M Oc&Oc.R ≠ 269·3°	Leading lights (≠ means "in line"; the firm line is the track to be followed; the light descriptions will be at the light stars or on the leading line, not usually both).	Occ.4s12M ☆ Occ.R. 4s10M Lights in line 269°18´ †	433.2 433.3 475.6
20.3	LdgOc.W&R ★	Leading lights on small-scale charts	Oc.W&R ★ 265°	433.1 475.6
21	Fl.G ☆ Fl.G ☆ 270° 2Fl.R ☆ 270°	Lights in line (for example, marking the sides of a channel)	Lights in line 092° Fl Fl †	433.4 475.6
22	Rear Lt or Upper Lt	Rear or upper light	Upr. †	470.7
23	Front Lt or Lower Lt	Front or lower light	Lr †	470.7

Direction Lights

30.1	☆ Dir 269° Fl(2)5s10m11M	Direction light with narrow sector and course to be followed, flanked by darkness or unintensified light	DirLt †	
30.2	Oc.12s6M ☆ Dir 299° Dir 255·5° ☆ Fl(2)15s11M	Direction light with course to be followed. Sector(s) uncharted	DirLt †	471.3 471.9 475 475.1 475.5 475.7
30.3	☆ F.G Al.WG Oc.W.4s Al.WR F.R DirWRG. 15-5M	Direction light with narrow fairway sector flanked by light sectors of different characters on standard charts		
30.4	☆ F.G Al.WG Oc.W.4s Al.WR F.R DirWRG. 15-5M #	Direction light with narrow fairway sector flanked by light sectors of different characters on multicoloured charts		
31	▶⊙ Dir 286°	Moiré effect light (day and night), variable arrow mark. Arrows show when course alteration needed		475.8

		Sector Lights		
40.1	Fl.WRG.4s21m 18-12M	*Sector light on standard charts*		470.4 475 475.1 475.2 475.5
40.2	Fl.WRG.4s21m 18-12M	*Sector light on multicoloured charts*		
41.1	Oc.WRG. 10-6M	*Sector lights on standard charts, the white sector limits marking the sides of the fairway*		470.4 475 475.1 475.5
41.2	Oc.WRG. 10-6M	*Sector lights on multicoloured charts, the white sector limits marking the sides of the fairway*		
42.1	Fl(3)10s62m25M F.R.55m12M	*Main light visible all-round with red subsidiary light seen over danger on standard charts*		471.8 475.4
42.2	Fl(3)10s62m25M F.R.55m12M	*Main light visible all-round with red subsidiary light seen over danger on multicoloured charts*		
43	Fl.5s41m30M	*All-round light with obscured sector*	Fl.5s41m30M	475.3
44	Iso.WRG	*Light with arc of visibility deliberately restricted*		475.1
45	Q.14m5M	*Light with faint sector*		475.3

46	Oc.R.8s R.5M R.9M R.5M / Oc.R.8s5M R R.Intens R	Light with intensified sector		475.2
a		Light with unintensified sector	Oc.R.8s R.5M R.2M / Oc.R.8s5/2M R R.Unintens	

Lights with limited Times of Exhibition

50	F.R(occas)	Lights exhibited only when specially needed (e.g. for fishing vessels, ferries) and some private lights	† (fishg.) † (Priv.) † (occasl.)	473.2
51	Fl.10s40m27M (F.37m11M Day)	Daytime light (charted only where the character shown by day differs from that shown at night)	† Fl.10s40m27M (F.37m11M by Day)	473.4
52	Q.WRG.5m10-3M (Fl.5s Fog)	Fog light (exhibited only in fog, or character changes in fog)	† Q.WRG.5m10-3M Fl.5s (in Fog)	473.5
53	† Fl.5s(U)	Unwatched (unmanned) light with no standby or emergency arrangements		473.1
54	# (temp)	Temporary	† (temp) † (tempy.)	473.6
55	(exting)	Extinguished	† (extingd.)	473.7

Special Lights   Flare Stack (at Sea) → L   Flare Stack (on Land) → E   Signal Stations → T

60	AeroAl.Fl.WG.7·5s11M		Aeronautical light (may be unreliable)		476.1
61.1	† AeroF.R.353m11M RADIO MAST (353)		Air obstruction light of high intensity		476.2
61.2	(89) (R Lts)		Air obstruction lights (e.g. on radio mast)	† (Red Lt.)	
62	Fog Det Lt		Fog detector light		477
63	◁▷	◁▷	(illum)	Floodlit, floodlighting of a structure	† (lit)	478.2
64	F Iso F.R Strip light			Strip light		478.5
65	# (priv)		Private light other than one exhibited occasionally	# ⊙ Y.Lt # ⊙ R.Lt † (Priv)	473.2
66	(sync) or (sync)		Synchronized (synchronous or sequential)		478.3

	IALA Maritime Buoyage System, which includes Beacons → Q 130		Buoys and Beacons	

			General	
1	–o–	*Position of buoy or beacon*		455.3 460.1 462.1

	Abbreviations for colours (lights) → P 11		Colour of Buoys and Beacons	
2	G B G G G	*Single colour; green (G) and black (B)*	† B G	450 450.1 450.2 450.3 464 464.1 464.2 464.3
3	R R Y Y Or R	*Single colour other than green and black: red (R), yellow (Y), orange (Or)*	† R Y Or	
4	BY GRG BRB	*Multiple colours in horizontal bands: the colour sequence is from top to bottom*	† BW RW BR BW	
5	RW RW BuY RW	*Multiple colours in vertical or diagonal stripes; the darker colour is given first. In these examples, red(R), white(W), blue (Bu), yellow (Y) & black(B)*	† RW BR BW BW	
6		*Retroreflecting material may be fitted to some unlit marks. Charts do not usually show it. Black bands will appear dark blue under a spotlight*	† Refl	
a		*Single colour other than green and black (non-IALA system: white (W) grey (Gy), blue (Bu))*	† W Gy Bu (non-IALA) W (non-IALA) Gy (non-IALA) Bu	464
b		*Wreck buoy (not used in the IALA System)*	† G G G G	
c		*Chequered*	† BR BW RW BW	

	Marks with Fog Signals → R		Lighted Marks	
7	Fl.G G Fl.R R	*Lighted marks on standard charts (examples)*	†	457.1 466 466.1
8	Fl.R R Iso RW Fl.G G	*Lighted marks on multicoloured charts (examples)*		

	For Application of Topmarks within the IALA System → Q 130	*Radar reflector* → S	Topmarks and Radar Reflectors	
9	(topmark symbols)	*IALA System buoy topmarks (beacon topmarks shown upright)*	Non-IALA System # (topmark symbols) etc.	463 463.1
10	Name 2 R	*Beacon with topmark, colour, radar reflector and designation (example)*	'2' R No.2 R Ra.Refl "2" †	450 455.2 455.7 455.8
11	Name 3 G	*Buoy with topmark, colour, radar reflector and designation (example). Radar reflectors are not generally charted on IALA System buoys*	'3' G No.3 Ra.Refl "No.3" †	460.3 460.6 465.1 465.2

173

Q Buoys, Beacons

Buoys	Features Common to Beacons and Buoys → Q 1-11

Shapes

20	⌂	▲	Conical buoy, nun buoy, ogival buoy	†	⌂ ▲ ◢ etc.	462.2
21	⌂	◣	Can buoy, cylindrical buoy	†	⌂ ▨ ■ etc.	462.3
22	⌂	⌂	Spherical buoy	†	⌂ ⊜ ● etc.	462.4
23	⟁ ⟁ ⟁		Pillar buoy	†	⟁ ⟙	462.5
24	⌶		Spar buoy, spindle buoy	†	⌶ ⌶ ⌶ ⌶ ⌶	462.6
25	⌂	⬭	Barrel buoy, tun buoy			462.7
26	⊏⊐		Superbuoy. Superbuoys are very large buoys, e.g. an aid to navigation mounted on a circular hull of about 5m diameter. Oil or gas installation buoys (L16) and ODAS buoys (Q58), of similar size, may be shown by variations of the superbuoy symbol	†	⊏⊐	445.4 460.4 462.9 474

Light Vessels and Minor Light Floats

30	⚓ Fl.G.3s G Name		Light float as part of IALA System			462.8
31	† ⚓ Fl.10s		Light float not part of IALA System	†	⚑ ⚑ ⚑ ⚑ ⚑ B R B	462.8
32	⚓		Light vessel	†	⚓ ⚓ LtV ⚓	

Mooring Buoys

		Oil or Gas Installation Buoy → L		Visitors' (Small Craft) Mooring → U		
40	◣ # ⌂ # ⌂ # ◣		Mooring buoy		# ■ # ⌂ # ⌂ † ⌂	431.5
41	◣ Fl.Y.2·5s		Lighted mooring buoy (example)			431.5 466.1 466.2 466.3 466.4
42	(trot diagram with buoys ①②)		Trot, mooring buoys with ground tackle and berth numbers	†	(trot diagram with buoys ①②)	431.6
43	◣ ∿∿∿∿∿∿		Mooring buoy with telephonic communications			431.5
44	⌐ Small Craft Moorings ⌐		Numerous moorings (example)			431.7
45	◉		Visitor's mooring			431.5

			Special Purpose Buoys	
	The symbols shown below are examples: shapes of buoys may differ; lateral or cardinal buoys may be used in some situations; the use of the 'X' topmark is optional.			
50	DZ	Firing danger area (Danger Zone) buoy		441.2
51	Target	Target		
52	Marker Ship	Marker Ship		
53	Barge	Barge		
54	DG	Degaussing Range buoy		448.2
55	Cable	Cable buoy	Cable †	443.6
56		Spoil ground buoy		446.3
57		Buoy marking outfall		444.2
58	ODAS ODAS	Ocean (or Oceanographic) Data Acquisition System (ODAS) buoy, Data collection buoy	ODAS †	448.3 460.4 462.9
60		Seaplane anchorage buoy		
61		Buoy marking traffic separation scheme		
62		Buoy marking recreation zone		
63	Al.Oc.BuY.3s BuY	Emergency wreck marking buoy		461.3 463.1 466.2
70	(priv)	Buoy privately maintained (example)		
71	(Apr-Oct))	Seasonal buoy (the example shows a yellow spherical buoy on station between April and October)	# (1.4 - 15.10) (occas)	460.5
d		Racing mark	#	

Q Buoys, Beacons

Beacons	Lighted Beacons → P	Features Common to Beacons and Buoys → Q 1-11

General

80	⊥ ⊙ Bn	Beacon in general, characteristics unknown or chart scale too small to show	⊥ #	455.5
81	⊥ BW	Beacon with colour, no distinctive topmark (example)		455.4 456 459.2
82	⊡ R ⬗ BY ⬤ BRB	Beacon with colour and topmark (examples)	W B R BW etc. †	455.4 456 459.2 463 463.1
83	⌿ BRB	Beacon on submerged rock (topmark and colours as appropriate)	⬤ BRB #	455.6
e		Beacon which does not conform with the IALA system	⊥ (non-IALA) W	

Minor Impermanent Marks usually in Drying Areas (Lateral Mark for Minor Channel)

Minor Pile → F

	PORT HAND	STARBOARD HAND			
90	⊥		Pole	⟟ †	456.1
91	Y	↑	Perch, withy	⟟ †	456.1
92	⚥ †	⚥ †	Withy		

Minor Marks, usually on Land

Landmarks → E

100	⚙	Cairn	⊙ Cairn †	456.2
101	▫ Mk	Coloured or white mark (the colour may be indicated)		456.2
102.1	⬆ RW ⬆ †	Coloured topmark (colour known or unknown) with function of a beacon	▫ R ⬆ G	
102.2	⬆ RW ⬆ RW †	Painted boards with function of leading beacons		

Beacon Towers

110	⬠ R ▪ G ⬠ R ⬠ G ⬠ BY ⬤ BRB	Beacon towers without and with topmarks and colours (examples)	† ⬠ Bn Tower † ⬠ Bn Tr etc.	456.4
111	⬚	Lattice beacon #		456.4

Leading Lines, Clearing Lines → M			Special Purpose Beacons	

Note: Topmarks and colours are shown where scale permits				

120	270°	Leading beacons (the firm line is the track to be followed)	Bn Bn Ldg Bns 270° †	458
121	270°	Beacons marking a clearing line or transit	Bn Bn Lts in line 270° †	458
122	Measured Distance 1852m 088·5°-268·5°	Beacons marking measured distance with quoted bearings. The track is shown as a firm line if it is to be followed precisely		458
123		Cable landing beacon (example)		443.5 458
124	Ref #	Ref	Refuge beacon	456.4
125			Firing practice area beacons	
126			Notice board	NB	456.2

130	IALA Maritime Buoyage System	IALA International Association of Marine Aids to Navigation and Lighthouse Authorities	NP735

Where in force, the IALA System applies to all fixed and floating marks except landfall lights, leading lights and marks, sectored lights and major floating lights.

The standard buoy shapes are cylindrical (can) 🛢, conical 🔺, spherical 🔵, pillar 🗼, and spar 🚩, but variations may occur, for example: minor light floats ⛴. In the illustrations on the next page, only the standard buoy shapes are used. In the case of fixed beacons (lit or unlit) only the shape of the topmark is of navigational significance.

IALA Buoyage Regions A and B
There are two international buoyage regions where lateral marks differ.
Region A is primarily comprised of the waters surrounding Greenland, Europe, Africa, Australia and Asia (except for Japan, the Republic of Korea and the Philippines).
Region B is primarily comprised of the waters surrounding North and South America, Japan, the Republic of Korea and the Philippines (see illustration).

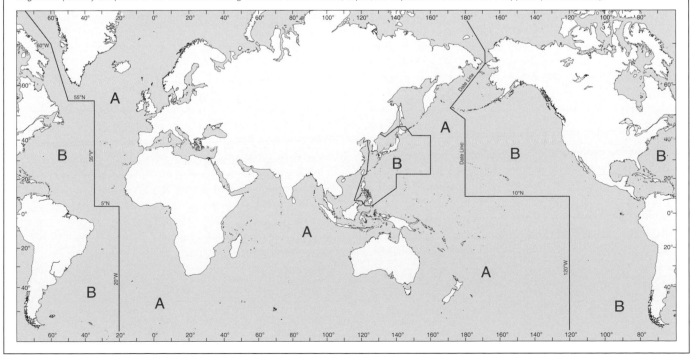

Q Buoys, Beacons

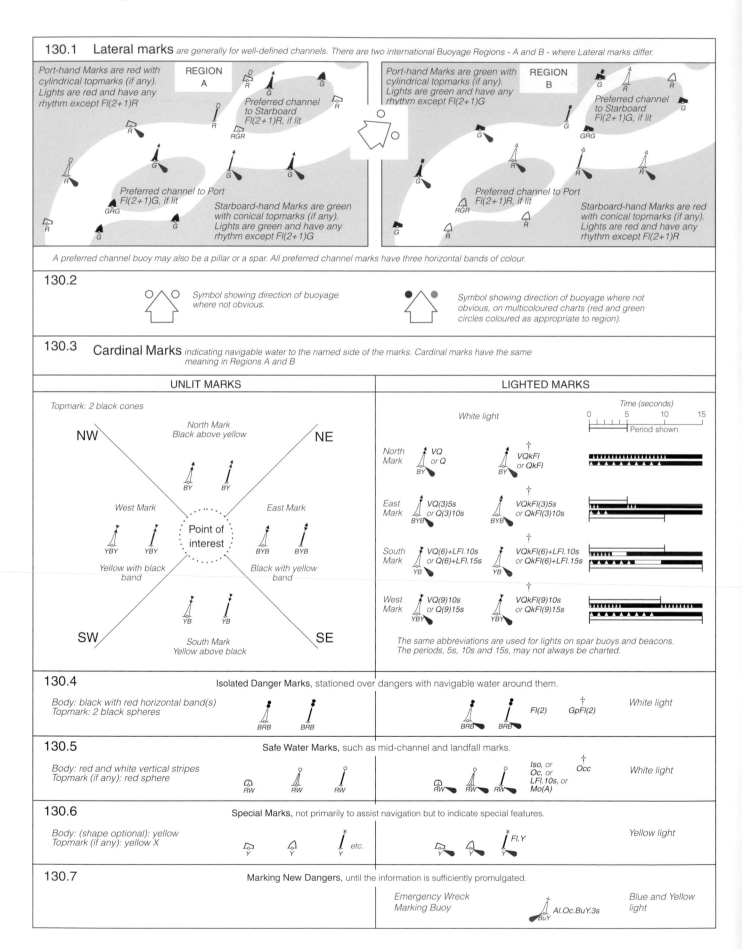

130.1 Lateral marks *are generally for well-defined channels. There are two international Buoyage Regions - A and B - where Lateral marks differ.*

REGION A

Port-hand Marks are red with cylindrical topmarks (if any). Lights are red and have any rhythm except Fl(2+1)R

Preferred channel to Starboard Fl(2+1)R, if lit

Preferred channel to Port Fl(2+1)G, if lit

Starboard-hand Marks are green with conical topmarks (if any). Lights are green and have any rhythm except Fl(2+1)G

REGION B

Port-hand Marks are green with cylindrical topmarks (if any). Lights are green and have any rhythm except Fl(2+1)G

Preferred channel to Starboard Fl(2+1)G, if lit

Preferred channel to Port Fl(2+1)R, if lit

Starboard-hand Marks are red with conical topmarks (if any). Lights are red and have any rhythm except Fl(2+1)R

A preferred channel buoy may also be a pillar or a spar. All preferred channel marks have three horizontal bands of colour.

130.2

Symbol showing direction of buoyage where not obvious.

Symbol showing direction of buoyage where not obvious, on multicoloured charts (red and green circles coloured as appropriate to region).

130.3 Cardinal Marks *indicating navigable water to the named side of the marks. Cardinal marks have the same meaning in Regions A and B*

UNLIT MARKS	LIGHTED MARKS

Topmark: 2 black cones

North Mark — Black above yellow — **NW** — **NE**

West Mark — Point of interest — East Mark

Yellow with black band — Black with yellow band

SW — South Mark — Yellow above black — **SE**

White light

Time (seconds) 0 5 10 15 — Period shown

		White light			
North Mark	VQ or Q	BY	† VQkFl or QkFl	BY	
East Mark	VQ(3)5s or Q(3)10s	BYB	† VQkFl(3)5s or QkFl(3)10s	BYB	
South Mark	VQ(6)+LFl.10s or Q(6)+LFl.15s	YB	† VQkFl(6)+LFl.10s or QkFl(6)+LFl.15s	YB	
West Mark	VQ(9)10s or Q(9)15s	YBY	† VQkFl(9)10s or QkFl(9)15s	YBY	

The same abbreviations are used for lights on spar buoys and beacons. The periods, 5s, 10s and 15s, may not always be charted.

130.4 Isolated Danger Marks, stationed over dangers with navigable water around them.

Body: black with red horizontal band(s)
Topmark: 2 black spheres

BRB BRB   BRB BRB   Fl(2)   † GpFl(2)   White light

130.5 Safe Water Marks, such as mid-channel and landfall marks.

Body: red and white vertical stripes
Topmark (if any): red sphere

RW RW RW   RW RW RW   Iso, or Oc, or LFl.10s, or Mo(A)   † Occ   White light

130.6 Special Marks, not primarily to assist navigation but to indicate special features.

Body: (shape optional): yellow
Topmark (if any): yellow X

Y Y Y etc.   Y   Fl.Y   Yellow light

130.7 Marking New Dangers, until the information is sufficiently promulgated.

Emergency Wreck Marking Buoy   Al.Oc.BuY.3s   BuY   Blue and Yellow light

	Fog Detector Light → P	Fog Light → P			General
1	(((°	Position of fog signal. Type of fog signal not stated		† Fog Sig	451 451.2 452.8

Types of Fog Signals, Abbreviations

10	Explos	Explosive	† Gun	452.1
11	Dia	Diaphone		452.2
12	Siren	Siren		452.3
13	Horn	Horn (nautophone, reed, tyfon)	† Nauto † E.F. Horn † Tyfon † Reed	452.4
14	Bell	Bell		452.5
15	Whis	Whistle		452.6
16	Gong	Gong		452.7

Examples of Fog Signal Descriptions

20	Fl.3s70m29M Siren Mo(N)60s	Siren at a lighthouse, giving a long blast followed by a short one (N), repeated every 60 seconds		452.3 453.3
21	Bell	Wave-actuated bell buoy. The provision of a legend indicating number of emissions, and sometimes the period, distinguishes automatic bell or whistle buoys from those actuated by waves		452.5 453 454.1
22	Q(6)+LFl.15s Horn(1)15sWhis YB	Light buoy, with horn giving a single blast every 15 seconds, in conjunction with a wave-actuated whistle	Reserve fog signals are fitted to certain buoys Only those actuated by waves are charted	452.4 453.1 454.3

‡ The Fog Signal symbol (R1) is usually omitted when associated with another navigation aid (e.g. light or buoy) when a description of the signal is given

S Radar, Radio, Satellite Navigation Systems

Radar	*Radar Structures Forming Landmarks* → E		*Radar Surveillance Systems* → M		
1	⊙ Ra	Coast radar station providing range and bearing from station on request			485.1
2	⊙ Ramark	Ramark, radar beacon transmitting continuously			486.1
3.1	† ⊙ Racon(Z) (3cm)	Radar transponder beacon, with morse identification, responding within the 3cm (X) band	† ⊙ Racon(Z)		
3.2	† ⊙ Racon(Z) (10cm)	Radar transponder beacon, with morse identification, responding within the 10cm (S) band			486.2 486.3
3.3	⊙ Racon(Z)	Radar transponder beacon, with morse identification	⊙ Racon(Z) (3 & 10cm) †		
3.4	Racon Obscd ⊙ Racon(P)	Radar transponder beacon with sector of obscured reception			486.4
3.4	Racon(Z) ⊙ Racon(Z)	Radar transponder beacon with sector of reception			486.4
3.5	Racon ⊙---⊙ Racon — Racons ≠ 270°	Leading radar transponder beacons (≠ and ≠ mean "in line")			486.5 433.3
3.5	Racon ★---★ Racon — Lts ≠ 270° Racons ≠ 270°	Leading radar transponder beacons coincident with leading lights			486.5 433.3
3.6	Racon ⌂ Racon	Radar transponder beacons on floating marks (examples)			486.2
4	⋏	Radar reflector (not usually charted on IALA System buoys and buoyant beacons)	Ra.Refl. †		455.8 459.2 465
5	⋏	Radar conspicuous feature	Ra conspic †		485.2

					Radio
	Radio Structures Forming Landmarks → E		*Radio Reporting (Calling-in or Way) Points* → M		
10	† Name RC	Non-directional marine or aeromarine radiobeacon			480 481.1 480.1
11	† RD 269·5° RD	Directional radiobeacon with bearing line	† Dir.Ro.Bn Dir.Ro.Bn 269°30′		480 481.2
	† Lts≠ 270° RD RD 270°	Directional radiobeacon coincident with leading lights (‡ means "in line")			
12	† RW	Rotating pattern radiobeacon			480 481.1
13	† Consol	Consol beacon			480
14	† RG	Radio direction-finding station	† Ro.D.F		483
15	† R	Coast radio station providing QTG service	† Ro.		480 484
16	† Aero RC	Aeronautical radiobeacon			480 482
17.1	AIS	Automatic Identification System transmitter			489.1
17.2	AIS AIS	Automatic Identification System transmitters on floating marks (examples)			489.1
18.1	V-AIS	Virtual AIS aid to navigation with no known IALA-defined function. Other carriers may be used			489.2
18.2	V-AIS V-AIS V-AIS V-AIS	Virtual AIS aid to navigation with IALA cardinal mark function			
18.3	V-AIS V-AIS	Virtual AIS aid to navigation with IALA lateral mark function			
18.4	V-AIS	Virtual AIS aid to navigation with IALA isolated danger mark function			
18.5	V-AIS	Virtual AIS aid to navigation with IALA safe water mark function			
18.6	V-AIS	Virtual AIS aid to navigation with IALA special mark function			
18.7	V-AIS	Virtual AIS aid to navigation with IALA new danger mark function			

					Satellite Navigation Systems
50	WGS WGS72 WGS84	World Geodetic System, 1972 or 1984			201
Note:	*A note may be shown to indicate the shifts of latitude and longitude, to one, two or three decimal places of a minute, depending on the scale of the chart, which should be made to satellite-derived positions (which are referred to WGS84) to relate them to the chart.*				202
51	# DGPS	Station providing Differential Global Positioning System corrections			481.4

T Services

Pilotage

1.1	◉	Pilot boarding place, position of pilot cruising vessel	† *Pilots*	† Pilots	
1.2	◉ *Name*	Pilot boarding place, position of pilot cruising vessel, with name (e.g. District, Port)			491.1 491.2
1.3	◉ *Note*	Pilot boarding place, position of pilot cruising vessel, with note (e.g. Tanker, Disembarkation)			
1.4	◉ *H*	Pilots transferred by helicopter			
2	† ■ Pilot lookout	Pilot office with Pilot lookout, Pilot lookout station			
3	■ Pilots	Pilot office			491.3
4	Port Name (Pilots)	Port with pilotage service (boarding place not shown)			491.4

Coastguard, Rescue

10	■CG	⊙CG	♇CG	Coastguard station	■CGFS	492 492.1 492.2
11	■CG♦	⊙CG♦	♇CG♦	Coastguard station with Rescue station	■CGFS♦	493.3
12		♦		Rescue station, Lifeboat station, Rocket station	† LB	493 493.1
13		🛶♦	♦	Lifeboat lying at a mooring		493.2
14	Ref		*Ref*	Refuge for mariners		456.4

182

		Stations			
20	⊙SS	Signal station in general	† Sig Sta	† Sig Stn	494
21	⊙SS(INT)	Signal station showing International Port Traffic Signals			495.4
22	⊙SS(Traffic)	Traffic signal station, Port entry and departure signals			495.1
23	⊙SS(Port Control)	Port control signal station			495.1
24	⊙SS(Lock)	Lock signal station			495.2
25.1	⊙SS(Bridge)	Bridge passage signal station			495.3
25.2 †	F Traffic Sig	Bridge lights including traffic signals			
26	⊙SS	Distress signal station			
27	⊙SS	Telegraph station			
28	⊙SS(Storm)	Storm signal station	† Storm Sig	† Stm. Sig. Stn.	497.1
29	⊙SS(Weather)	Weather signal station, Wind signal station			497.1
30	⊙SS(Ice)	Ice signal station			497.1
31	⊙SS(Time)	Time signal station			
32.1	‡	Tide scale or gauge	⊙Tide gauge		496.1
32.2	⊙Tide gauge	Automatically recording tide gauge			
33	⊙SS(Tide)	Tide signal station			496.2
34	⊙SS(Stream)	Tidal stream signal station			496.3
35	⊙SS(Danger)	Danger signal station			497.2
36	⊙SS(Firing)	Firing practice signal station			497.2

U   Small Craft (Leisure) Facilities

Small Craft (Leisure) Facilities	Transport Features, Bridges →D Public Buildings, Cranes →F		Pilots, Coastguard, Rescue, Signal Stations →T		
1.1	⚓	Boat harbour, Marina			320.2
1.2	⛵	Yacht berths without facilities			
2	Ⓥ	Visitors´ berth			321.8
3		Visitors´ mooring	†	Ⓥ	
4	▶	Yacht club, Sailing club			320.2
5		Public slipway	†	◣	
7		Public landing, Steps, Ladder	†	⌐	
10		Public house, Inn	†	🍺	
11		Restaurant	†	✕	
17		Water tap	†	�🚰	
18		Fuel station (Petrol, Diesel)	†	⛽	
19		Electricity	†	⚡	

22		Laundrette	†	⊙	
23		Public toilets	†	**wc**	
24		Post box	†	🗑	
25		Public telephone	†	✆	
26		Refuse bin	†	🗑	
27		Public car park	†	**P**	
28		Parking for boats and trailers	†	⊥	
29		Caravan site	†	🚐	
30		Camping site	†	⛺	

32

MARINA FACILITIES

HARBOUR / MARINA FACILITIES	Diesel	Petrol	Bottled Gas	Electricity	Holding Tank Disposal	Scrubbing Berth	Repairs	Crane/Boat Hoist	Launching Slip	Pontoon Berthing	Swinging Moorings	Chandlery	Laundrette	VHF Radio Channels	Showers	Telephone Area Code	Telephone Number	Fax Number
FALMOUTH - Falmouth Visitors Yacht Haven							●			●	●	●		●	●	12	+44 (0) 1326	312285	211352
- Mylor Yacht Harbour	●	●	●	●	●	●	●	●	●	●	●	●	●	●	●	80/M	+44 (0) 1326	372121	372120
† HELFORD - Helford Moorings Officer											●			●	●	-	+44 (0) 1326	250749	-

Marina Facilities are no longer inserted on ADMIRALTY charts. Users are recommended to contact the marina, or visit their website, for the latest information. Contact details are given on some ADMIRALTY charts.

Abbreviations of Principal Non-English Terms

Glossaries of non-English terms will be found in the volumes of ADMIRALTY Sailing Directions.

On metric ADMIRALTY charts, non-English terms are generally given in full wherever space and information permits. Where abbreviations are used on metric charts they accord with the following list, apart from those on charts published before 1980 where full stops are omitted. Obsolescent forms of abbreviations may also be found on these charts and on reproductions of other nations' charts.

CURRENT FORM	OBSOLESCENT FORM(S)	TERM	ENGLISH MEANING
ALBANIAN			
	K	Kodër, Kodra	*Hill*
ARABIC			
	Djeb, Dj	Djebel	*Mountain, Hill*
Geb.	G	Gebel	*Mountain, Hill*
J.	Jab, J^l	Jabal, Jibāl, Jebel	*Mountain(s), Hill(s)*
Jaz.	Jaz^t	Jazīrat, Jazā'ir Jazīreh	*Island(s), Peninsula*
Jeb.	J, J^l	Jebel	*Mountain, Hill*
Jez.	Jez^t	Jezīrat	*Island, Peninsula*
Kh.	K	Khawr, Khōr	*Inlet, Channel*
	Si, S^i	Sidi	*Tomb*
W.		Wād, Wādi	*Valley, River, River bed*
CHINESE			
Chg.	Ch^g	Chiang	*River, Shoal, Harbour, Inlet, Channel, Sound*
DANISH			
B.		Bugt	*Bay, Bight*
Bk.	B^k	Banke	*Bank*
Fj.	F^d	Fjord	*Inlet*
Gr.	Grd, Gr^d, G^d	Grund	*Shoal*
H.	Hm, H^m, Hne, H^{ne}	Holm, Holmene	*Islet(s)*
Hd.	H^d	Hoved	*Headland*
Hn.	H^n	Havn, Havnen	*Harbour*
Ll.		Lille	*Little*
N.		Nord, Nordre	*North, Northern*
Ø.		Øst, Østre	*East, Eastern*
Øy.	Øne, $Ø^{ne}$, Öne, $Ö^{ne}$	Øyane, Øyene, Öyane Öyene	*Islands*
Pt.	P^t	Pynt	*Point*
S.		Sønder, Søndre	*South, Southern*
Sd.	S^d	Sund, Sundet	*Sound*
Sk.	Skr, Sk^r	Skær, Skjær	*Rock above water*
St.		Stor	*Great*
V.		Vest, Vestre	*West*
DUTCH			
B.	B^i	Baai	*Bay*
Bg.	B^g	Berg	*Mountain*
Bk.	B^k	Bank	*Bank*
Eil.	Eiln, Eil^n	Eiland, Eilanden	*Island(s)*
G.		Golf	*Gulf*
	Gt, Grt, G^t, Gr^t	Groot, Groote	*Great*
H.		Hoek	*Cape, Hook*
Pt.	P^t	Punt	*Point*
R.		Rivier	*River*
Rf.	R^f	Rif	*Reef*
Str.	Stn, St^r, St^n	Straat, Straten	*Strait(s)*
FINNISH			
K.		Kari, Kallio, Kivi	*Rock, Reef*
Lu.		Luoto, Luodet	*Rock(s)*
Ma.		Matala	*Shoal*
	P	Pieni, Pikku	*Small*
Sa.	S^a	Saari, Saaret	*Island(s)*
Tr.	T^r	Torni	*Tower*
FRENCH			
A.	A^e	Anse	*Inlet*
B.	B^e	Baie	*Bay*
	Bas. Bsse	Basse	*Shoal*
Bc.	B^c	Banc	*Bank*
	Bssn, Bn, B^n	Bassin	*Basin*
C.		Cap	*Cape*
Cal.	Ch^{al}, Chen	Chenal	*Channel*
Ch.	Chap, $Chap^e$	Chapelle	*Chapel*
Chât.	$Chât^u$, Ch^{au}	Château	*Castle*

CURRENT FORM	OBSOLESCENT FORM(S)	TERM	ENGLISH MEANING
FRENCH *(continued)*			
F.	F^l	Fleuve	*Large river*
Ft.	F^t	Fort	*Fort*
G.		Golfe	*Gulf*
	Gd, G^d, Gde, G^{de}	Grand, Grande	*Great*
Ht.Fd.	H.F., Ht fd, $H^t fd$, H^t fond	Haut-fond	*Shoal*
Î.	I, I^t	Île, Îles, Îlot	*Island(s), Islet*
L.		Lac	*Lake*
	Mn, M^{in}	Moulin	*Mill*
Mlg.	Mge, M^{age}, Mou	Mouillage	*Anchorage*
Mt.	M^t	Mont	*Mount, Mountain*
	N.D.	Notre Dame	*Our Lady*
P.		Port	*Port*
	Pet, P^{it}, P^{ite}, P^t	Petit, Petite	*Small*
Pit.	Pn, P^{on}	Piton	*Peak*
Pl.		Plage	*Beach*
Plat.	Pla, $Plat^u$	Plateau	*Tableland, Sunken flat*
Pte.	P^{te}	Pointe	*Point*
Qu.	Q	Quai	*Quay*
R.	Rau, Riv, R^{au}	Rivière, Ruisseau	*River, Stream*
	Rav, R^{ne}	Ravine	*Ravine*
Rf.		Récif	*Reef*
Roc.	Re, R^e, Rer, R^{er}	Roche, Rocher	*Rock*
S.	St, S^t, Ste, S^{te}	Saint, Sainte	*Saint, Holy*
	Som.	Sommet	*Summit*
Tr.	T^r	Tour	*Tower*
	Vi, V^x	Vieux, Vieil, Vielle	*Old*
GAELIC			
Bo.		Bogha	*Below water rock*
Eil.	E, En, E^n	Eilean, Eileanan	*Island(s), Islet(s)*
Ru.	R^u	Rubha	*Point*
Sg.	Sgr, Sg^r	Sgeir	*Rock*
GERMAN			
B.		Bucht	*Bay*
Bg.	B^g	Berg	*Mountain*
Gr.	Grd, Gr^d, G^d	Grund	*Shoal*
Hn.	H^n	Hafen	*Harbour*
K.		Kap	*Cape*
Rf.	R^f	Riff	*Reef*
	Schl	Schloss	*Castle*
GREEK			
Ág., Ag.	Áy., Ay.	Ágios, Ágia	*Saint, Holy*
Ágk.	Ang.	Agkáli	*Bight, Open bay*
Ágky.	Angir., Ang	Agkyrovólio	*Anchorage*
Ák., Ak.		Ákra, Akrotírio	*Cape*
Kól.	Kol	Kólpos	*Gulf*
Lim.		Limín, Liménas	*Harbour*
N.		Nísos, Nísoi	*Island(s)*
N.	N	Nisída, Nisídes	*Islet(s)*
Ó.	O	Órmos	*Bay*
Or.		Ormískos	*Cove*
Ór.	Or	Óros, Óroi	*Mountain(s)*
Pot.		Potamós	*River*
	Prof	Profítis	*Prophet*
Sk.		Skópelos, Skópeloi	*Reef(s), Drying rock(s)*
Vrach.	Vrak	Vrachonisída, Vrachonisídes	*Rocky islets*
Vrach.	Vrák	Vráchos, Vráchol	*Rock(s)*
Ýf.	Íf.	Ýfalos, Ýfaloi	*Reef(s)*
ICELANDIC			
Fj.	Fjr, F^{dr}	Fjörður	*Fjord*
Gr.		Grunn	*Shoal*

Abbreviations of Principal Non-English Terms

INDONESIAN and MALAY

CURRENT FORM	OBSOLESCENT FORM(S)	TERM	ENGLISH MEANING
A.		Air, Ajer, Ayer	Stream
B.	Bu, B^u	Batu	Rock
Bat.	Btg, B^{tg}	Batang	River
	Bdr, B^{dr}	Bandar, Bendar	Port
	Br, B^r	Besar	Great
Buk.	Bt, B^t	Bukit	Hill
G.	Gg, G^g	Gosong, Gosung, Gusong, Gusung	Shoal, Reef, Islet
Gun.	Gg, G^g	Gunong, Gunung	Mountain
K.	Ki, K^i	Kali	River
K.	Kr	Kroeng, Krueng	River
Kam.	Kg, K^g	Kampong, Kampung	Village
Kar.	Kg, K^g	Karang	Coral reef, Reef
Kep.	Kpn, K^{pn}	Kepulauan	Archipelago
Kl.	K^l	Kachil, Kechil, Ketjil, Kecil	Small
Ku.	Kla, K^{la}	Kuala	River mouth
Lab.	Labn, Lab^n	Labuan, Labuhan	Anchorage, Harbour
Mu.	Ma, M^a	Muara	River mouth
P.	Pu, P^u, P^o	Pulau, Pulu, Pulo	Island
Peg.		Pegunungan	Mountain range
Pel.	Pln, P^{ln}	Pelabuan, Pelabuhan	Roadstead, Anchorage
P.–P.	P.P.	Pulau-pulau	Group of islands
	Prt, P^{rt}	Parit	Stream, Canal, Ditch
S.	Si, S^i	Sungai, Sungei	River
Sel.	Slt, S^{lt}	Selat	Strait
T.	Tg, T^g	Tandjong, Tandjung, Tanjong, Tanjung, Tanjing	Cape
Tel.	Tal, Tk, T^k	Taluk, Telok, Teluk	Bay
U.	Ug, U^g	Udjung, Ujung	Cape
W.		Wai	River

ITALIAN

CURRENT FORM	OBSOLESCENT FORM(S)	TERM	ENGLISH MEANING
Anc.		Ancoraggio	Anchorage
B.		Baia	Bay
Banch.	Bna, B^{na}	Banchina	Quay
Bco.	B^{co}	Banco	Bank
C.		Capo	Cape
Cal.		Calata	Wharf
Can.		Canale	Channel
Cas.		Castello	Castle
F.		Fiume	River
Fte.	F^{te}	Forte	Fort
G.		Golfo	Gulf
	Gde, G^{de}	Grande	Great
I.	I^a, I^e	Isola, Isole	Island(s)
I.	I^{to}, I^{ti}	Isolotto, Isolotti	Islet(s)
L.		Lago	Lake
Lag.	La, L^e	Laguna	Lagoon
	Mda, Mad, Mad^a, Mad^{na}	Madonna	Our Lady
Mte.	M^{te}	Monte	Mount, Mountain
P.	Pto, P^{to}	Porto	Port
P.	Portlo, Port^{lo}	Porticciolo	Small port
Pco.	P^{co}	Picco	Peak
Pog.	Pgio, P^{gio}	Poggio	Mound, Small hill
Pta.	P^{ta}	Punta	Point, Summit
	Pte, P^{te}	Ponte	Bridge
	Pzo, P^{zo}	Pizzo	Peak
S.	Sto, S^{to}, Sta, Sta	San, Santo, Santa	Saint, Holy
S.	SS, S.S.	Santi	Saints
Scog.	Sço, Sçi, Sc, Sc^i	Scoglio, Scogli	Rock(s), Reef(s)
Scog.	Sc, Scra	Scogliera	Ridge of rocks, Breakwater
Sec.	Se	Secca, Secche	Shoal(s)
	T, T^{te}	Torrente	Intermittent stream
Tr.	Tre, T^{re}	Torre	Tower
	Va, V^{la}	Villa	Villa

JAPANESE

CURRENT FORM	OBSOLESCENT FORM(S)	TERM	ENGLISH MEANING
B.	B^a	Bana	Cape, Point
By.	Bi, B^i	Byōchi	Anchorage
	D^e	Dake	Mountain, Hill
G.	G^a	Gawa	River
H.	Ha, H^a	Hana	Cape, Point
Hak.	Hi, H^i	Hakuchi	Roadstead

JAPANESE (continued)

CURRENT FORM	OBSOLESCENT FORM(S)	TERM	ENGLISH MEANING
J.	J^a	Jima	Island
K.	Ka, K^a	Kawa	River
	Kaik, Ko, K^o	Kaikyō	Strait
M.	Mki, M^{ki}, M^i	Misaki	Cape
	Ma, M^a	Mura	Village
	Mi, M^i	Machi	Town
S.	Si, S^i	Saki	Cape, Point
Sh.	Sa, S^a	Shima	Island
	Sn, S^n	San	Mountain
	So, S^o	Seto	Strait
Su.	Sdo, S^{do}	Suidë	Channel
	Te, T^e	Take	Hill, Mountain
	Ya, Y^a	Yama	Mountain
Z.	Z^i	Zaki	Cape, Point
	Z^n	Zan	Mountain

MALAY (see INDONESIAN)

NORWEGIAN

CURRENT FORM	OBSOLESCENT FORM(S)	TERM	ENGLISH MEANING
B.	B, B^{kt}	Bukt, Bukta	Bay, Bight
Bg.	B^g	Berg, Bierg, Bjerg	Mountain, Hill
Fd.	F^d, Fj	Fjord, Fjorden	Fjord
Fjel.	Fj	Fjell, Fjellet, Fjeld, Fjeldet	Mountain
Fl.	Flne, Fl^{ne}	Flu, Flua, Fluen, Fluane, Fluene	Below water rock(s)
Gr.	Grne, Gr^{ne}	Grunn, Grunnen, Grunnane	Shoal(s)
H.	Hm, H^m, Hne, H^{ne}	Holm, Holmen, Holmane	Islet(s)
Hn.	H^n	Hamn, Havn	Harbour
in.	In^r, I	Indre, Inre, Inste	Inner
L.		Lille, Liten, Litla, Litle	Little
Lag.	La, L^a	Laguna	Lagoon
N.		Nord, Nordre	North, Northern
Ø.	Ö	Øst, Østre, Öst, Östre	East, Eastern
Od.	O	Odde, Odden	Point
Øy.	Ø, Ö, O	Øy, Øya, Öy, Öya	Island
Øy.	Øne, Ø^{ne}, Öne, Ö^{ne}	Øyane, Øyene, Öyane, Öyene	Islands
Pt.	P^t	Pynt, Pynten	Point
S.		Syd, Søre, Søndre	South, Southern
Sd.	S^d	Sund, Sundet	Sound
Sk.	Skr, Sk^r	Skjær, Skjer, Skjeret	Rock above water
Sk.	Skne, Sk^{ne}	Skjerane, Skjærane	Rocks above water
St.		Stor, Stora, Store	Great
Tar.	Tn, T^n	Taren	Below water rock
V.		Vest, Vestre	West
Vag.	Vg, V^g	Våg, Vågen	Bay, Cove
	Vd, V^d	Vand	Lake
Vik.	Vk, V^k	Vik, Vika, Viken	Bay, Inlet
	Vn, V^n	Vann, Vatn	Lake
Y.	Y^t	Ytre, Ytter, Yttre	Outer

PERSIAN

CURRENT FORM	OBSOLESCENT FORM(S)	TERM	ENGLISH MEANING
B.		Bandar	Harbour
Jab.		Jabal	Mountain, Hill
Jaz.	Jazh, Jaz^h	Jazīreh	Island, Peninsula
Kh.	K	Khowr	Inlet, Channel
R.		Rūd	River

POLISH

CURRENT FORM	OBSOLESCENT FORM(S)	TERM	ENGLISH MEANING
Jez.		Jezioro	Lake
Kan.		Kanal	Channel
Miel.		Mielizna	Shoal
R.		Rzeka	River
W.	Wys, Wa, W^a	Wyspa	Island
Zat.		Zatoka	Gulf, Bay

PORTUGUESE

CURRENT FORM	OBSOLESCENT FORM(S)	TERM	ENGLISH MEANING
Anc.		Ancoradouro	Anchorage
Arq.	Arqu^o	Arquipélago	Archipelago
B.		Baía	Bay
Bco.	B^{co}	Banco	Bank
Bxo.	Ba, B^{xo}, Bxa, B^{xa}	Baixo, Baixa, Baixia, Baixio	Shoal
Co.	C.	Cabo	Cape

Abbreviations of Principal Non-English Terms

PORTUGUESE (continued)

CURRENT FORM	OBSOLESCENT FORM(S)	TERM	ENGLISH MEANING
Can.		Canal	Channel
Ens.	Ens^a	Enseada	Bay, Creek
Est.	Est^o	Esteiro	Creek, Inlet
Estr.		Estreito	Strait
Estu.	Est, Est^o	Estuario	Estuary
	Fte, F^{te}	Forte	Fort
	Fte, Ftza, F^{tza}	Fortaleza	Fortress
Fund.		Fundeadouro	Anchorage
G.		Golfo	Gulf
	Gde, G^{de}	Grande	Great
I.		Ilhéu, Ilhéus, Ilhota	Islet(s)
I.		Ilha, Ilhas	Island(s)
	L.	Lago	Lake
	L.	Lagoa	Small lake, Marsh
La.	Le, L^e	Laje	Flat-topped rock
Lag.	La, L^a	Laguna	Lagoon
	Mol, Me, M^e	Molhe	Mole
	Mor, Mo, M^o	Morro	Headland, Hill
Mt.	M^{te}, Mte	Monte, Montanha	Mount, Mountain
NS.	Na.Sa, N^{a}S^a	Nosso Senhor, Nossa Senhora	Our Lord, Our Lady
P.	Pto, P^{to}	Porto	Port
	Pal, Pals, Pal^s	Palheiros	Fishing village
	Par, Pel, P^{el}	Parcel	Shoal, Reef
Pass.	Pas	Passagem, Passo	Passage, Pass
	Pco, P^{co}, P^o	Pico	Peak
	Pda, P^{da}	Pedra	Rock
	Peq	Pequeno, Pequena	Small
	Pr, Pa, P^a	Praia	Beach
Pta.	P^{ta}	Ponta	Point
	Queb.	Quebrada, Quebrado	Cut, Ravine
Rch.		Riacho, Ribeira, Ribeirão	Creek, Stream, River
Rf.		Recife	Reef
Ro.	R	Rio	River
Roc.	Ra, R^a	Rocha, Rochedo	Rock
S.	Sto, S^{to}, Sta, Sta	São, Santo, Santa	Saint, Holy
Sa.	Sa, S^a, Sr	Serra, Cordilheira	Mountain range
	Va, V^a	Vila	Town, Village, Villa

ROMANIAN

CURRENT FORM	OBSOLESCENT FORM(S)	TERM	ENGLISH MEANING
A.		Ansă, Ansa	Cove
B.		Baie, Baia	Bay
Br.		Braţ, Braţul, Braţu	Branch, Arm (of the sea)
C.		Cap, Capul, Capu	Cape
Di., D-le.		Deal, Dealul, Dealuri, Dealurile	Hill(s)
Fd.mic		Fund mic	Shoal
I.		Insulă, Insula	Island
L.		Lac, Lacul, Lacu	Lake
Mt., M-ţii.		Munte, Muntele, Munţi, Muntii	Mountain, Mounts
O.		Ostrov, Ostrovul, Ostrovu	Island
S.		Stîncă, Stînca	Rock
Sf.		Sfînt, Sfîntu, Sfîntul, Sfînta	Saint, Holy
Str.		Strîmtoare, Strîmtoarea	Pass, Strait

RUSSIAN

CURRENT FORM	OBSOLESCENT FORM(S)	TERM	ENGLISH MEANING
B		Bukhta	Bay, Inlet
b-ka.	Bka, B^{ka}, Bki, B^{ki}, Bk	Banka, Banki	Bank(s)
Bol.		Bol'shoy, Bol'shaya, Bol'shoye	Great, Large
Gb.	G, Ga, G^a	Guba	Gulf, Bay, Inlet
G.		Gora	Mountain, Hill
Gav.	G	Gavan'	Harbour, Basin
Kam.		Kamen'	Rock
M.		Mys	Cape, Headland
	Mal	Malyy, Malaya, Maloye	Little
O.	O^{va}	Ostrov, Ostrova	Island(s)
Oz.		Ozero	Lake
P–ov.	Pol^{ov}, P^{ov}, Pol	Poluostrov	Peninsula
Pr.	Prv, Pr^v	Proliv	Channel, Strait
R.		Reka	River
Zal.		Zaliv	Gulf, Bay

SPANISH

CURRENT FORM	OBSOLESCENT FORM(S)	TERM	ENGLISH MEANING
	A, Arro, Arr^o	Arroyo	Stream
Arch.	Arch^o	Archipiélago	Archipelago
Arrf.	Arr^e, Arr^{fe}, Arr	Arrecife	Reef
Ba.	B^a	Bahía	Bay
	Bo, B^o	Bajo	Shoal
	Bco, B^{co}	Banco	Bank
Br.	Bzo, B^{zo}	Rompientes	Breakers
C.		Cabo	Cape
	Cal, Cta	Caleta	Cove
Can.		Canal	Channel
	Cer, Co, C^o	Cerro	Hill
Cre.		Cumbre, Cima	Summit
	Cy	Cayo	Cay, Key
Ens.	Ens^a	Ensenada	Cove
	Est, Est^o	Estero	Creek, Inlet
Estr.		Estrecho	Strait
Estu.	Est, Est^o	Estuario	Estuary
Fond.	Fond^o	Fondeadero	Anchorage
	Fte, F^{te}	Fuerte	Fort
G.		Golfo	Gulf
	Gde, G^{de}	Grande	Great
I, Is	I^a	Isla, Islas	Island(s)
	I, I^{te}	Islote, Isleta	Islet
	L.	Lago	Lake
	Lag, La, L^a	Laguna	Lagoon
	Mor, Mo, M^o	Morro	Headland, Hill
	Mte, M^{te}	Monte	Mount, Mountain
	Mu, Me, M^e, Mlle	Muelle	Mole
	Na. Sa, N^{a}S^a	Nuestra Señora	Our Lady
	P, Pto, P^{to}	Puerto	Port
	Pco, P^{co}, P^o	Pico	Peak
	Pda, P^{da}	Piedra	Rock
Pen.	Pen^{la}	Península	Peninsula
	Peq	Pequeño, Pequeña	Small
	Pl, Pa, P^a	Playa	Beach
Prom.	Prom^{to}	Promontorio	Promontory
	Pta, P^{ta}	Punta	Point
	Queb.	Quebrada	Cut, Ravine
	R.	Río	River
	Rga.	Restinga	Shoal, Sandbank
	Roc, Ra, R^a	Roca	Rock
S.	Sn, S^n, Sto, S^{to}, Sta, Sta	San, Santo, Santa	Saint, Holy
	Sr, Sa, S^a	Sierra	Mountain range
	Surg, Surgo, Surg^o	Surgidero	Anchorage, Roadstead
Tr.	Te, T^{re}	Torre	Tower
	Va, V^a	Villa	Villa, Small town

SWEDISH

CURRENT FORM	OBSOLESCENT FORM(S)	TERM	ENGLISH MEANING
B.		Bukt	Bay, Bight
Bg.	Bgt, B^g	Berg, Berget	Mountain
	Bk, B^k	Bank	Bank
Fj.	F^d	Fjärd, Fjord	Fjord
	Gla, G^{la}	Gamla	Old
Gr.	Grn, Grd, Grd, G^d	Grund	Shoal
H.	Hm, H^m	Holme, Holmarna	Islet
	Hd, H^d	Huvud	Headland
	Hn, H^n	Hamn, Hamnen	Harbour
I.		Inre	Inner
L.		Lilla, Liten	Little, Small
N.		Nord, Norra	North, Northern
Ö.		Öst, Östra	East, Eastern
S.		Syd, Södra	South, Southern
Sk.	Sk^r	Skär, Skäret, Skären	Rock above water
St.		Stor	Great, Large
V.		Väst, Västra	West, Western
Y.	Y^t	Yttre	Outer

THAI

CURRENT FORM	OBSOLESCENT FORM(S)	TERM	ENGLISH MEANING
Kh.		Khao	Hill, Mountain
L.	Lm, L^m	Laem	Cape, Point
M.N.		Mae Nam	River

TURKISH

CURRENT FORM	OBSOLESCENT FORM(S)	TERM	ENGLISH MEANING
Ad.		Ada, Adası	Island
Aşp		Takimadalar	Archipelago
Adc.	Ad	Adacık	Islet
Boğ.		Boğaz, Boğazı	Strait
Br.	Bn, Bu	Burun, Burnu	Point, Cape
Ç.	Ça	Çay, Çayı	Stream, River

Abbreviations of Principal Non-English Terms

CURRENT FORM	OBSOLESCENT FORM(S)	TERM	ENGLISH MEANING
TURKISH *(continued)*			
	Da	Dağ, Dağı	*Mountain*
D.	De	Dere, Deresi	*Valley, Stream*
Dz.		Deniz	*Sea*
G.		Göl, Gölü	*Lake*
Isk.		İskele, İskelesi	*Jetty*
Kf. Krf.		Körfez, Körfezi	*Gulf*
Ky.	Kyl.	Kaya, Kayası	*Rock*
Lim. Lm.	Li	Liman, Limanı	*Harbour*
N.		Nehir, Nehri, Irmak, Irmağı	*River*
T.	Te, Tᵉ	Tepe, Tepesi	*Hill, Peak*
Yad.		Yarımada, Yarımadası	*Peninsula*

CURRENT FORM	OBSOLESCENT FORM(S)	TERM	ENGLISH MEANING
Languages of the former YUGOSLAVIA			
Br.		Brdo, Brda	*Mountain(s)*
Gr.		Greben, Grebeni	*Rock, Reef, Cliff, Ridge*
Hr.		Hrid, Hridi	*Rock*
L.		Luka	*Harbour, Port*
M.		Mali, Mala, Malo, Malen	*Small*
O.		Otočić, Otočići	*Islet(s)*
O.		Otok, Otoci	*Island(s)*
Pl.		Pličina	*Shoal*
Pr.		Prolaz	*Passage*
S.	Sv	Sveti, Sveta, Sveto	*Saint, Holy*
Šk.		Školj, Školjić	*Island, Reef*
U.		Uvala, Uvalica	*Inlet*
V.		Veli, Vela, Velo, Velik, Veliki, Velika, Veliko	*Great*
Z.	Zal	Zaliv, Zaljev, Zaton	*Gulf, Bay*

Abbreviations of Principal English Terms

(Note: INT abbreviations are in bold type)

CURRENT FORM	OBSOLESCENT FORM(S)	TERM	REFERENCES
abt	ab^t	About	—
Accom		Accommodation Vessel	L17
Aero		Aeronautical	P 60, 61
AIS		Automatic Identification System	S 17, 18
	Al	Algae	J s
Al.	Alt	Alternating light	P 10.11
ALC		Articulated Loading Column	L 12
ALL		ADMIRALTY List of Lights and Fog Signals	—
ALRS		ADMIRALTY List of Radio Signals	—
Am		Amber	P 11.8
Anch.	Anch^e	Anchorage	—
	Anct, Anc^t	Ancient	—
ANM		Annual Summary of ADMIRALTY Notices to Mariners	—
Annly	Ann^{ly}	Annually	—
Appr.	Apprs, Appr^s	Approaches	—
approx	Approx	Approximate	—
Apr		April	—
Arch.	Archo, Arch^o	Archipelago	—
ASD		ADMIRALTY Sailing Directions	—
ASL		Archipelagic Sea Lane	M17
	Astr, Astrl, Astr^l	Astronomical	—
ATBA		Area to be Avoided	M14, 29
ATT		ADMIRALTY Tide Tables	—
Aug		August	—
Aus		Australia	—
Ave	Ave^e	Avenue	—
B.		Bay	—
B	bl, blk	**Black**	J af, **Q 2**
	Ba	Basalt	J h
	Batt, Baty, Bat^y	Battery	E 34.3
Bk.	B^k	Bank	—
bk	brk	**Broken**	J 33
Bldg	B^{ldg}	Building	D 5
	BM, B.M.	Bench Mark	B 23
Bn, Bns		Beacon(s)	M 1–2, P 4–5, Q 80–81
BnTr	Bn Tower	Beacon Tower	P 3, Q 110
Bo		Boulders	J 9.2
Bol	Boll.	Bollard	F a
Br		Breakers	K 17
	br	Brown	J ak
Bu	Bl, Bl., b	Blue	J ag, P 11.4, Q a
C.		Cape	—
c		Coarse	J 32
ca	cal	Calcareous	J 38
CALM		Catenary Anchor Leg Mooring	L 16
Cas	Cas.	Castle	E 34.2
	Cath, Cath.	Cathedral	E 10.1
Cb		Cobbles	J 8
cd		Candela	B 54
CD		Chart Datum	H 1
	Cemy, Cem^y	Cemetery	E 19
CG	C.G.	Coastguard station	T 10–11
Ch	Ch.	Church, chapel	E 10.1
	ch, choc	Chocolate	J al
Chan.		Channel	—
Chem		Chemical	L 40
	chk, Ck	Chalk	J e
Chy	Chy^y	Chimney	E 22
	cin, Cn	Cinders	J m
cm	cm.	Centimetre(s)	B 43
Co	crl	Coral	J 10, K 16
	Col	Column, pillar, obelisk	E 24
	conspic	Conspicuous	E 2
const	constn, constr^n	Construction	F 32
cov	cov.	Covers	K b
Cr.		Creek	—
Cup	Cup.	Cupola	E 10.4
Cy	cl	Clay	J 3
	(D)	Doubtful	—
	d	Dark	J ao
Dec		December	—
decrg	decr^g	Decreasing	B 64
dest	destd, Dest^d	Destroyed	—
Det		(see Fog Det Lt)	—
DG	D. G.	Degaussing	N 25, Q 54
DGPS		Differential Global Positioning System	S 51
	Di, di	Diatoms	J v
Dia		Diaphone	R 11
Dir	Dir^n	Direction	—
Dir	Dir Lt	Direction light	P 30–31
Discol	Discol^d	Discoloured water	K d
discont	discontd, discont^d	Discontinued	—
dist	Dist	Distant	—
Dk	D^k	Dock	—
dm	dm.	Decimetre(s)	B 42
Dn, Dns	D^n	Dolphin(s)	F 20
dr	dr., Dr.	Dries	K a
DW		Deep-water, Deep-draught	M 27, N 12.4
dwt		Deadweight tonnage	—
DZ		Danger Zone	Q 50
E	E.	East	B 10
ED	(ED), (E.D.)	Existence doubtful	I 1
EEZ		Exclusive Economic Zone	N 47
	E.F. Horn	Electric fog horn	R 13
Ent.	Entce, Ent^{ce}	Entrance	—
	Equin^l	Equinoctial	—
ESSA		Environmentally Sensitive Sea Area	N 22
Est.	Est^y	Estuary	—
	Estab^t	Establishment	—
	ev.	Every	—
exper	experl, Exper^l	Experimental	—
explos	explos.	Explosive	R 10
(exting)	(exting^d)	Extinguished	P 55
f		Fine	J 30
F		Fixed	P 10.1
FAD		Fish Aggregating Device	—
F Racon		Fixed frequency radar transponder beacon	S 3.4
Feb		February	—
FFL		Fixed and flashing light	P 10.10
Fj.	Fd, F^d	Fjord	—
	(fish^g)	Fishing light	P 50
Fl	fl.	Flashing	P 10.4
	Fl., fl	Flood	—
Fla		Flare stack (at sea)	L 11
	Fm, F^m	Farm	—
fm, fms	f^m, f^{ms}	Fathom, fathoms	B 48
Fog Det Lt		Fog detector light	P 62
	Fog Sig.	Fog signal station	R 1
	Fog W/T	Radio fog signal	—
FPSO		Floating Production and Storage Offtake Vessel	L17
	Fr, for	Foraminifera	J t
FS	F.S.	Flagstaff, Flagpole	E 27
FSO		Floating Storage and Offtake Vessel	L17
FSU		Floating Support Unit	L17
	Ft, F^t	Fort	E 34.2
ft	f^t	Foot, feet	B 47, P 13
G	g	Gravel	J 6
G	gn	Green	J ah, P 11.3, Q 2
G.		Gulf	—
	ga, glac	Glacial	J ac
	Gc	Glauconite	J o
	Gd, grd	Ground	J a
	Gl, gl	Globigerina	J u
	Govt Ho, Gov^t Ho	Government House	—
GNSS		Global Navigation Satellite System	—
Gp.		Group (of islands)	—
	GpFl, Gp.Fl.	Group-flashing	P 10.4
	GpOcc, Gp.Occ.	Group-occulting	P 10.2
GPS		Global Positioning System	—
grt		Gross Register Tonnage	—
	Gt, Grt, G^t, Gr^t	Great	—
	G.T.S.	Great Trigonometrical Survey Station (India)	—
	Gy, gy	Grey	J am, Q a
GT		Gross Tonnage	—
h		Hard	J 39
	H, H.	Headway	D 20, D 26–27
H		Helicopter transfer (Pilots)	T 1.4
h	h., H.	Hour	B 49
HAT		Highest Astronomical Tide	H 3
Hd.	H^d	Headland	—
Hn.	H^n	Haven	—

Abbreviations of Principal English Terms

(Note: INT abbreviations are in bold type)

CURRENT FORM	OBSOLESCENT FORM(S)	TERM	REFERENCES
Ho		House	—
(hor)	(hor^l)	Horizontally disposed	P 15
Hosp	Hospl, Hosp^l	Hospital	F 62.2
Hr.	H^r	Harbour	—
	Hr, H^r	Higher	—
Hr Mr		Harbour Master	F 60
	Ht, H^t	Height	—
HW	H.W.	High Water	H a
	H.W.F. & C.	High Water Full and Change	—
	H.W.O.S.	High Water Ordinary Springs	—
I.	It	Island, islet	—
IALA		International Association of Lighthouse Authorities	Q 130
IHO		International Hydrographic Organization	—
(illum)	Illum., (lit)	Illuminated	P 63
IMO		International Maritime Organization	—
	in., ins.	Inch, inches	—
incrg	incr^g	Increasing	B 65
INT		International	A 2, T 21
Intens	(intens)	Intensified	P 46
IQ	IntQkFl, Int.Qk.Fl.	Interrupted quick-flashing	P 10.6
	(irreg.)	Irregular	—
	ISLW, I.S.L.W.	Indian Spring Low Water	—
Iso		Isophase	P 10.3
	It	Islet	—
ITZ		Inshore Traffic Zone	—
IUQ		Interrupted ultra quick-flashing	P 10.8
IVQ	IntVQkFl, Int.V.Qk.Fl	Interrupted very quick-flashing	P 10.7
Jan		January	—
Jul		July	—
km	km.	Kilometre(s)	B 40
kn	kn.	Knot(s)	B 52, H 40 41
L.		Lake, Loch, Lough	—
	l	Large	J ab
Lag.	Lagn, l ag^n	Lagoon	—
LANBY		Large Automatic Navigational Buoy	P 6, Q 26
LASH		Lighter Aboard Ship	—
LAT		Lowest Astronomical Tide	H 2
Lat	Lat.	Latitude	B 1
	LB, L.B.	Lifeboat station	T 12
Ldg	L^{dg}	Leading	P 20.3
Le.	L^e	Ledge	—
LFl		Long-flashing	P 10.5
	Lit, Lit.	Little	—
	(lit)	Floodlit	P 63
LL	L.L.	List of Lights	—
Lndg	L^{dg}	Landing place	F 17
LNG		Liquefied Natural Gas	—
LOA		Length overall	—
LoLo		Load-on, Load-off	—
Long	Long.	Longitude	B 2
LPG		Liquefied Petroleum Gas	—
	Lr, L^r	Lower	P 23
	L.S.S.	Lifesaving station	—
Lt	L^t, lt	Light	J an, P 1
Lts		Lights	P 61.2
LtHo	L^t Ho	Lighthouse	P 1
Lt V	l^t V	Light vessel	Q 62
	Lv, lv	Lava	J i
LW	L.W.	Low Water	H b
	L.W.F. & C.	Low Water Full and Change	—
	L.W.O.S.	Low Water Ordinary Springs	—
M	m	Mud	J 2
M	M.	Sea or Nautical Mile(s)	B 45, P 14
m		Medium	J 31
m	m.	Metre(s)	B 41, P 13
	mad, Md	Madrepore	J g
Mag	Mag.	Magnetic	B 61
	Magz, Mag^z	Magazine	—
	man, Mn	Manganese	J n

CURRENT FORM	OBSOLESCENT FORM(S)	TERM	REFERENCES
Mar		March	—
MHLW	M.H.L.W.	Mean Higher Low Water	H 14
MHW		Mean High Water	H 5
MHWN	M.H.W.N.	Mean High Water Neaps	H 11
MHWS	M.H.W.S.	Mean High Water Springs	H 9
	Mid, Mid.	Middle	—
min	min., m.	Minute(s) of time	B 50
Mk		Mark	Q 101
	Ml, ml	Marl	J c
MLHW	M.L.H.W.	Mean Lower High Water	H 15
MLLW	M.L.L.W.	Mean Lower Low Water	H 12
MLW		Mean Low Water	H 4
MLWN	M.L.W.N.	Mean Low Water Neaps	H 10
MLWS	M.L.W.S.	Mean Low Water Springs	H 8
mm	mm.	Millimetre(s)	B 44
Mo		Morse code	P 10.9, R 20
Mon	Mont, Mon^t	Monument	E 24
	Mony, Mon^y	Monastery	—
	Ms, mus	Mussels	J 9
MR		Marine reserve	N 22
MRCC		Maritime Rescue and Coordination Centre	—
MSL	M.S.L.	Mean Sea Level	H 6
Mt.	M^t	Mountain, mount	—
Mth.	Mth	Mouth	—
MTL	M.T.L.	Mean Tide Level	H c
N	N.	North	B 9
	Nauto	Nautophone	R 13
NB	N.B.	Notice Board	Q 126
NE	N.E.	North-east	B 13
NM	N.M.	Notice(s) to Mariners	—
n mile		International Nautical Mile	B 45
No	N^o	Number	N 12.2
Nov		November	—
Np	Np.	Neap Tides	H 17
nrt		Nett register tonnage	—
NT		Net Tonnage	—
NW	N.W	North-west	B 15
NZ		New Zealand	—
	Obs Spot, Obsn Spot, Obs^n Spot	Observation Spot	B 21
Obscd	Obsc^d	Obscured	P 43
Obstn	Obst^n	Obstruction	K 40-43, L 43
	Obsy, Obs^y	Observatory	—
Oc	Occ, Occ.	Occulting	P 10.2
(occas)	(occas^l)	Occasional	P 50
Oct		October	—
OD	O.D.	Ordnance Datum	H d
ODAS		Ocean Data Acquisition System	Q 58
	Off, Off.	Office	—
Or	Or.	Orange	P 11.7, Q 3
	ord.	Ordinary	—
	Oy, oys	Oysters	J p
	Oz, oz	Ooze	J b
P	peb	Pebbles	J 7
P.		Port	—
(P)		Preliminary (NM)	—
PA	(PA), (P.A.)	Position approximate	B 7
Pag	Pag.	Pagoda	E 13
Pass.		Passage	—
PD	(PD), (P.D.)	Position doubtful	B 8
Pen.	Penla, Pen^{la}	Peninsula	—
Pk.	P^k	Peak	—
	Pm, pum	Pumice	J j
PO	P.O.	Post Office	F 63
	Po, pol	Polyzoa	J y
pos	posn, pos^n	Position	—
(priv)	priv., (Priv.)	Private	P 65, Q 70
	Prod Well	Production Well	L 20
prohib	Prohib^d	Prohibited	—
proj	projd, Proj^d	Projected	—
prom	promt, Prom^t	Prominent	—
Prom.	Promy, Prom^y	Promontory	—
	(prov), (prov^l)	Provisional	—
PSSA		Particularly Sensitive Sea Area	N 22
Pt.	P^t	Point	—
	Pt, pt	Pteropods	J x
Pyl		Pylon	D 26

Abbreviations of Principal English Terms

(Note: INT abbreviations are in bold type)

CURRENT FORM	OBSOLESCENT FORM(S)	TERM	REFERENCES
Q	QkFl, Qk.Fl.	Quick-flashing	P 10.6
	Q^r	Quarter	—
	Qz, qrtz	Quartz	J f
R	rd	Red	J aj, P 11.2, Q 3
R.		River	—
R	r	Rock	J 9.1, K 15
	R, R^o	Coast Radio Station providing QTG service	S 15
Ra		Radar, Coast Radar Station	M 31–32, S 1
	Ra (conspic), Ra. (conspic)	Radar conspicuous object	S 5
	Ra. Refl.	Radar Reflector	Q 10–11, S 4
Racon		Radar Transponder Beacon	S 3.1–3.6
	rad, Rd	Radiolaria	J w
Ramark		Radar Beacon	S 2
	RC	Non-directional Radio-beacon	S 10
	RD, Dir.Ro.Bn	Directional Radiobeacon	S 11
Rds.	R^{ds}	Roads, Roadstead	—
Ref		Refuge	Q 124, T 14
Refl	Refl.	Retroreflecting material	Q 6
	Rem^{ble}	Remarkable	—
Rep	Repd, Rep^d	Reported	I 3
Rf.	R^f	Reef	—
RG	R^o D.F.	Radio Direction-Finding Station	S 14
Rk.	R^k	Rock	—
(R Lts)	(Red Lts)	Air Obstruction Lights (low intensity)	P 61.2
	Rly, Ry, R^y	Railway	D 13
	R^o B^n	Radiobeacon in general	S 10
RoRo	Ro-Ro	Roll-on Roll-off ferry terminal	F 50
	R.S.	Rocket station	—
Ru, (ru)	Ru.	Ruins, (ruined)	D 8, E 25.2, F 33
	RW	Rotating Pattern Radiobeacon	S 12
S.	St, S^t	Saint	—
S	s	Sand	J 1
S	S.	South	B 11
s	sec, sec.	Second(s) of time	B 51, P 12
SALM		Single Anchor Leg Mooring	L 12
SBM		Single Buoy Mooring	L 16
SC	S.C.	Sailing Club	U 4
	Sc, sc	Scoriæ	J l
Sc	Sc.	Scanner	E 30.3
Sch	Sch.	School	—
SD	S.D.	Sailing Directions	—
SD		Sounding of doubtful depth	I 2
Sd.	S^d	Sound	—
SE	S.E.	South-east	B 14
	Sem, Sem.	Semaphore	—
Sep		September	—
sf	stf	Stiff	J 36
Sh	sh	Shells	J 11
Sh.		Shoal	—
Si		Silt	J 4
	Sig, Sig.	Signal	R 1, T 25.2
	sk, spk	Speckled	J ad
	sm	Small	J aa
SMt	SM^t	Seamount	—
	Sn, shin	Shingle	J d
so	sft	Soft	J 35
Sp	Sp.	Spire	E 10.3
	Sp, sp	Sponge	J r
Sp	Sp, Spr.	Spring Tides	H 16
SPM		Single Point Mooring	L 12
SS	Sig Sta, Sig Stn	Signal Station	T 20–36
St	st	Stones	J 5
St	St.	Street	—
Sta	Sta., Stn, St^n	Station	D 13
	Stm.Sig.Stn.	Storm Signal Station	T 28
Str.		Strait	—
subm	submd, $Subm^d$	Submerged	—
SW	S.W.	South-west	B 16
SWOPS		Single Well Oil Production System	—
sy	stk	Sticky	J 34

CURRENT FORM	OBSOLESCENT FORM(S)	TERM	REFERENCES
	T, t	Tufa	J k
(T)		Temporary (NM)	—
t		Ton, tonne, tonnage	B 53, F 53
	t	Elevation of top of trees	C 14
Tel	Tel.	Telephone, Telegraph	—
(temp)	(tempy), ($temp^y$)	Temporary	P 54
Tr	T^r	Tower	E 10.2, E 20
TSS		Traffic Separation Scheme	—
TV Tr	T.V. T^r	Television Tower	E 28–29
	(U)	Unwatched, unmanned (light)	P 53
ULCC		Ultra Large Crude Carrier	—
uncov	uncov.	Uncovers	K c
unexam	unexamd. $unexam^d$	Unexamined	I a
Unintens		Unintensified	P a
	Up^r	Upper	P 22
UQ		Ultra quick-flashing	P 10.8
UTC		Co-ordinated Universal Time	—
UTM		Universal Transverse Mercator	—
v	vol	Volcanic	J 37
V-AIS		Virtual AIS aid to navigation	S18
	Va, V^a	Villa	—
Var	Var^n	Variation	B 60
	var	Varying	—
Vel	Vel.	Velocity	—
(vert)	($vert^l$)	Vertically disposed	P 15
Vi		Violet	P 11.5
	vis.	Visible	—
VLCC		Very Large Crude Carrier	—
Vol.		Volcano	—
VQ	VQkFl, V.Qk.Fl	Very quick-flashing	P 10.7
VTS		Vessel Traffic Service	—
W	W.	West	B 12
W	w	White	J ae, P 11.1, Q 130.5
Water Tr	Water T^r	Water tower	E 21
Wd	wd	Weed	J 13.1
Well		Wellhead	L 20, L 21
WGS		World Geodetic System	S 50
Whf	Wh^f	Wharf	F 13
Whis	Whis.	Whistle	R 15
Wk, Wks	W^k	Wreck(s)	K 20–30
	W/T	Radio (Wireless/Telegraphy)	—
Y	y	Yellow, amber, orange	J ai, P 11.8, Q 3
YC	Y.C.	Yacht Club	U 4
	y^d, y^{ds}	Yard(s)	—

Index

See also Abbreviations of principal English and non-English terms, including International Abbreviations.

Index

Index

CONTENTS KEY

Selection of Symbols

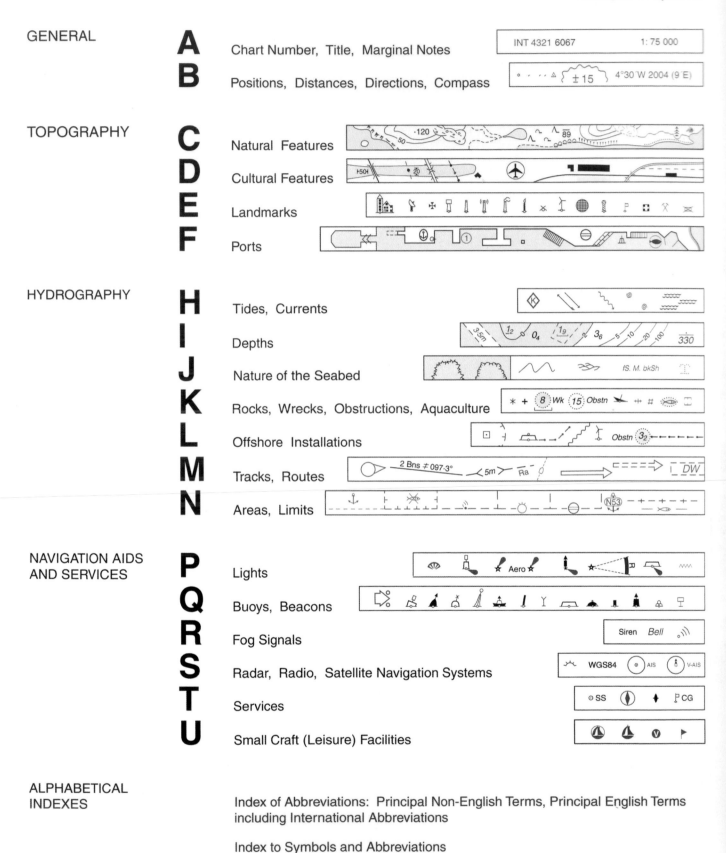

GENERAL	A	Chart Number, Title, Marginal Notes	INT 4321 6067 1 : 75 000
	B	Positions, Distances, Directions, Compass	± 15 4°30′W 2004 (9°E)
TOPOGRAPHY	C	Natural Features	
	D	Cultural Features	
	E	Landmarks	
	F	Ports	
HYDROGRAPHY	H	Tides, Currents	
	I	Depths	
	J	Nature of the Seabed	
	K	Rocks, Wrecks, Obstructions, Aquaculture	
	L	Offshore Installations	
	M	Tracks, Routes	
	N	Areas, Limits	
NAVIGATION AIDS AND SERVICES	P	Lights	
	Q	Buoys, Beacons	
	R	Fog Signals	
	S	Radar, Radio, Satellite Navigation Systems	
	T	Services	
	U	Small Craft (Leisure) Facilities	

ALPHABETICAL INDEXES

Index of Abbreviations: Principal Non-English Terms, Principal English Terms including International Abbreviations

Index to Symbols and Abbreviations

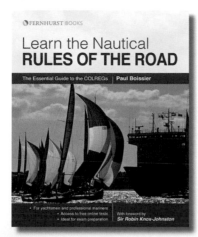

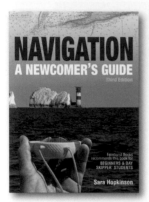
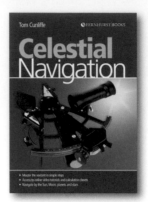
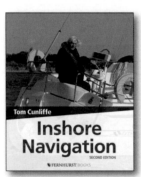

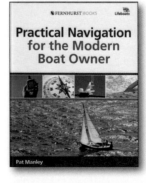